Ionic Reactions and Separations

EXPERIMENTS IN QUALITATIVE ANALYSIS

Ionic Reactions

and Separations

EXPERIMENTS IN QUALITATIVE ANALYSIS

EDWARD J. KING

Barnard College
Columbia University

Saunders College Publishing
Harcourt Brace Jovanovich College Publishers
Fort Worth   Philadelphia   San Diego
New York   Orlando   Austin   San Antonio
Toronto   Montreal   London   Sydney   Tokyo

Cover photo: Micrograph of xenon tetrafluoride crystals.
Argonne National Laboratory.

ISBN: 0-15-547041-8

Library of Congress Catalog Card Number: 72-86917

Printed in the United States of America

6 7 8 9 0   074   26 25 24 23 22

Preface

When my textbook *Qualitative Analysis and Electrolytic Solutions* was published in 1959, qualitative analysis was taught as a separate second-year course in the chemistry curriculum. Since then it has been almost universally incorporated into the general chemistry course. Today, as students encounter qualitative analysis in the laboratory, they are attending lectures and reading chemistry textbooks that give them the basic principles of the theory of solutions, electrolytes, and chemical equilibrium. Therefore, *Ionic Reactions and Separations*, while reviewing aspects of these subjects, seeks more expressly to show their specific applications to qualitative analysis—applications that are necessarily slighted in the general chemistry textbook.

Chapter 1 surveys various types of ionic reactions and shows how their extent under standard conditions can be judged from the magnitudes of K, $E°$, or $\Delta G°$. The sections on $E°$ and $\Delta G°$ can be omitted or deferred until these properties have been reached in the class lectures. Chapter 2 treats the important ionic separations, both by verbal description and by numerical calculation.

Chapter 3 presents the techniques of the analytical laboratory, which are then used in the procedures of Chapter 4 dealing with the detection of the anion. The principles introduced in Chapter 1 and applied in Chapter 4 are reviewed in Chapter 5 in a programmed format designed for self-study by the student. Procedures for detecting the cation are given in Chapter 6.

The analysis of simple substances containing only one cation and one anion presented in these chapters is an effective way to study ionic reactions. Therefore, courses in which only six weeks are assigned to qualitative analysis can use Chapter 1, parts of Chapter 2, and Chapters 3 through 6. Courses that have a full quarter or a term at their disposal can deal with the analysis of mixtures treated in Chapters 7 to 14.

The scheme of analysis for the cations given in *Qualitative Analysis and Electrolytic Solutions* has stood the test of time and has been modified in only a few details for its inclusion here. The original lengthy description of the chemistry of the ions has been replaced by a concise summary of the general characteristics of each group, but the discussion of the steps in the systematic analysis has been expanded so that the specific properties of the ions used in the analytical procedure are presented in some detail. Illustrative numerical calculations are also included to show the application of equilibrium principles to various steps in the procedure. In this book, the anion procedures have been simplified and the number of anions has been reduced from fifteen to eleven.

I take this opportunity to thank the many teachers and students whose comments and suggestions have been very helpful.

Edward J. King

To the Student

What is this white solid—baking powder or rat poison? Does that detergent contain phosphate? Can we detect manganese in this Etruscan sculpture and so prove that it is a forgery? Does this moon rock contain titanium? Such are the questions to which qualitative analysis can give a definite answer.

The true analyst, in contrast with the one-finger-in-the-book determinator, wants to know the principles on which his work is based. In qualitative analysis, he learns the practical implications and limitations of the theory of electrolytes. This is a mainstream of chemical interest that overflows into biochemistry, physiology, molecular biology, and geology. In qualitative analysis, for example, we use buffers to control separations between ions. The same principles of buffer action are important in regulating the acidity of body fluids. In qualitative analysis, we develop the principles governing the formation of precipitates. These principles apply also to formation of mollusk shells, to the deposition of calcium carbonate in the ocean or in caves, and to the creation of veins of sulfide minerals.

In the course of the laboratory work you, the analyst, will acquire experience with inorganic substances. The facts you learn may not seem important in themselves, but in the aggregate they constitute a reservoir of knowledge. With this experience you can independently interpret results and observations; without it you are forced to rely on the opinions and interpretations of others. You may ultimately forget much of this specific information about inorganic substances, especially if you study no more chemistry, but you will have experienced how scientists use knowledge gained by first-hand observations to shape their thought.

Scientific work is not required to be dull! Qualitative analysis will excite your sense of wonder. Add ammonia to a green solution of a nickel salt and watch it suddenly turn blue. Mix colorless solutions of an antimony salt and a sulfide and see a beautiful bright orange solid appear.

Qualitative analysis can be dull and routine if you choose to be a "cookbook" chemist. Or it can be enjoyable and satisfying if you keep your eyes open, develop your powers of deduction, and learn to be resourceful in dealing with the puzzling observation that does not jibe with your preconception or with some statement in this book.

Contents

TECHNIQUES, 45

ANALYSIS OF SIMPLE SUBSTANCES, 57

Six. Detection of the Cation   107

SYSTEMATIC ANALYSIS OF MIXTURES, 125

Seven. Cation Group 1.
The Hydrochloric Acid Group: Ag^+, Hg_2^{++}, Pb^{++} 129

Ten.   Cation Group 4.
The Ammonium Carbonate Group: Ca^{++}, Sr^{++}, Ba^{++} 183

Eleven.   Cation Group 5.
The Soluble Group: Na^+, K^+, Mg^{++}, NH_4^+ 198

APPENDIXES, 227

Principles

ONE

Ionic Reactions

1.1 Electrolyte Solutions

Electrolytes break up into ions in aqueous solution. Those that ionize completely, or nearly so, are called *strong electrolytes*. Their solutions have a comparatively large electrical conductivity. In dilute solution the common strong acids and bases and almost all salts are strong electrolytes (Table 1.1). *Weak electrolytes* are ionized to a small extent, and their solutions have low conductivities. The division of electrolytes into only two categories is necessarily an arbitrary one. Some electrolytes, such as sulfamic and iodic acids, calcium hydroxide, and cadmium chloride, fall in between.

The properties of an electrolyte solution, both physical and chemical, are the sum of the properties of its constituent ions and molecules, modified slightly by the strong forces that operate between ions even at a considerable distance from each other. An aqueous nickel(II) chloride solution, for example, is green because of the presence of hydrated nickel ions, and these same ions are responsible for the chemical reaction of the solution with ammonia to give blue $Ni(NH_3)_6^{++}$ ions. These are properties a nickel chloride solution has in common with other nickel(II) salts. It also reacts with silver nitrate to precipitate silver chloride, a common property of highly ionized chlorides.

3

Table 1.1

Summary of Strong and Weak Electrolytes

Rule	Exceptions
1. Most acids are weak electrolytes.	The common strong acids are HCl, HBr, HI, HNO_3, H_2SO_4,* $HClO_3$, and $HClO_4$.
2. Most bases are weak electrolytes.	The strong basic hydroxides are those of Li, Na, K, Rb, Cs, Ca, Sr, and Ba.*
3. Most salts are strong electrolytes.	The most important weakly ionized salt is $HgCl_2$.

* Only the loss of the first H^+ of H_2SO_4 or the first OH^- of $Ca(OH)_2$, $Sr(OH)_2$, and $Ba(OH)_2$ is complete.

1.2 Evidence for Chemical Reaction

There is no mistaking the occurrence of many reactions. The change in color from green to blue when ammonia is added to a nickel(II) salt is visual evidence of a reaction. When colorless lead nitrate and colorless potassium iodide solutions are mixed, we know a reaction occurs because we see the formation of flakes of yellow lead iodide. The evolution of gaseous ammonia by the action of sodium hydroxide on ammonium chloride is pungent notice of reaction.

When there is no simple visual or olfactory testimony of its occurrence, a reaction may be detected by some auxiliary test. The neutralization of lactic acid by sodium hydroxide, for example, produces no visible change, but if phenolphthalein is present the completion of the reaction is signalled by the appearance of the pink color of the indicator. We can follow this and many other reactions by using electric cells that respond to changes in concentration of H^+, Ag^+, or other ions. Conductivity measurements are sensitive to the presence of ions and can be used to chart a reaction's progress. There are many physical aids like these that will detect the occurrence of reaction by responding to the appearance or disappearance of ions.

According to any of these tests, no chemical reaction occurs when dilute solutions of potassium chloride and calcium nitrate are mixed. We might imagine a swapping of partners to give potassium nitrate and calcium chloride, but the properties of the mixture are simply those of potassium, nitrate, calcium, and chloride ions modified slightly by the long-range and nonspecific interactions between them. Both the original salts and the hypothetical products are soluble, strong electrolytes, so we might be tempted to write

$$2K^+ + 2Cl^- + Ca^{++} + 2NO_3^- \rightarrow 2K^+ + 2NO_3^- + Ca^{++} + 2Cl^-$$

But this only shows the same particles on both sides in different orders. No change has really occurred. No ions have been consumed. Nothing new has been produced. The two salts do not react in dilute solution.

1.3   Types of Reactions

It is helpful at this stage to have some general notion of the variety of possible reactions that may be encountered. No attempt has been made to make a comprehensive list. Details about many of the reactions are given later in the chapter.

Given a combination of two reagents in solution, can we make a stab at predicting plausible reactions? One approach is to try a swapping of partners. This may be expected to correspond to the reaction if one of the possible products is weakly ionized or insoluble.

Example 1   Mercury(II) nitrate and lithium chloride. The possible products are mercury(II) chloride and lithium nitrate, the first of which is one of the few weakly ionized salts (Table 1.1). We expect reaction to occur, removing Hg^{++} and Cl^- ions to give molecules of the weak salt.

Example 2   Ammonium chloride and sodium hydroxide. The possible products are ammonium hydroxide (aqueous ammonia), a weak base, and sodium chloride, a strong, soluble salt. The reaction involves the transfer of a hydrogen nucleus or proton from the ammonium ion, NH_4^+, to hydroxide ion to give weakly ionized ammonia and water.

Example 3   Hydrochloric acid and calcium acetate. The possible products are acetic acid, a weak electrolute, and calcium chloride, a strong, soluble salt. Proton transfer again occurs, this time from the hydrated hydrogen ion of the strong acid to the acetate ion.

Example 4   Zinc oxide and nitric acid. The possible products are water, a very weakly ionized substance, and the strong, soluble salt, zinc nitrate. Basic oxides like ZnO contain the oxide ion O^{--}, which accepts two hydrogen nuclei (protons) to form water.

Example 5   Silver nitrate and nickel(II) chloride. The possible products are silver chloride, almost insoluble in water, and nickel(II) nitrate, a strong, soluble salt. Silver and chloride ions are therefore removed from solution during the reaction to give the precipitate.

Other combinations cannot be handled by the device of partner swapping. Sometimes ancient rubrics are helpful.

Example 6   Sulfur dioxide and barium hydroxide. An old rule of thumb states that "acidic oxides react with basic hydroxides to give a salt and water." Which salt will it be? Remembering that sulfur dioxide reacts with water to give sulfurous acid, we expect to get the barium salt of this acid, barium sulfite. We could detect the occurrence of this reaction by the absorption of the gas by the basic hydroxide and the formation of a white precipitate of the salt.

Complex formation is often associated with some visible change: the solution turns a different color or a precipitate dissolves. As you gain familiarity with common complex ions you can anticipate their formation in certain situations.

Example 7 Aqueous ammonia and copper(II) nitrate. The deep blue $Cu(NH_3)_4{}^{++}$ ion is well known and may be expected to form upon addition of ammonia to a solution of this copper(II) salt because it is a weakly dissociated complex. Nitrate ions do not participate in the change.

Finally, a number of reactions may be spotted as involving electron transfer. These are oxidation–reduction or *redox* reactions. Simple displacement of one element by another is an example of such a change; other redox reactions are discussed in Secs. 1.11–1.13.

Example 8 Magnesium iodide and aqueous bromine. Iodide ions lose electrons and become elemental iodine, while bromine molecules gain electrons and become bromide ions. Magnesium ions are bystanders, taking no part in the redox reaction.

Chemists like to invent categories that are helpful in classifying reactions; for example, proton transfer and electron transfer reactions. Brønsted suggested that proton donors be called *acids* and proton acceptors, *bases*, so proton transfer reactions are often called acid–base reactions. Sulfur dioxide is usually described as an acidic oxide, yet it has no protons to donate.[1] G. N. Lewis suggested a generalization of the term acid to include electron-pair acceptors. Lewis bases are electron-pair donors as well as being proton acceptors. The reactions of Examples 6 and 7 are then classified as Lewis acid–base reactions:

$$\text{base} \quad + \quad \text{acid} \quad \rightarrow \text{combination}$$

$$\text{H—O:}^- \; + \; SO_2 \; \rightarrow HOSO_2{}^-$$

$$4H_3N\!: \; + \; Cu^{++} \; \rightarrow Cu(NH_3)_4{}^{++}$$

(The first reaction is followed by neutralization of $HOSO_2{}^-$ to $SO_3{}^{--}$ by more base.)

---

[1] Protons in the nuclei of atoms are not available for donation to a base. We mean by "proton" a hydrogen nucleus which in the acid is covalently bonded to some molecule or ion. Only the nucleus of hydrogen-1, the lightest isotope, is, strictly speaking, a proton.

1.4   Ionic Equations

Chemical equations show the substances that are involved in a reaction and their molar ratios, but they cannot represent all the complex aspects of electrolytic solutions or ionic reactions. It is helpful to adopt certain conventions about the representation of electrolytes in order that the equation may at least show the consumption or formation of ions. An equation such as

$$2KCl + Ca(NO_3)_2 \rightarrow CaCl_2 + 2KNO_3$$

gives no clue as to whether or not this is a bona fide reaction. We saw in Sec. 1.2 that in fact no such reaction occurs: all four substances are strong, soluble electrolytes, so that no new ions or molecules are produced.

We can avoid such pitfalls by following these conventions:

Rule 1.   Ionic formulas are written for strong electrolytes in solution. Examples: $Ca^{++} + 2NO_3^-$ and not $Ca(NO_3)_2$; $Ni^{++} + 2Cl^-$ and not $NiCl_2$ or $Ni^{++} + Cl_2^{--}$. We usually ignore hydration of ions and formation of ion pairs. Note that if you know the information in Table 1.1, you will be able to recognize strong electrolytes.

Rule 2.   Molecular formulas are written for all other substances: elements, gases, precipitates, and weak electrolytes or nonelectrolytes in solution. Examples: $I_2(s)$, $CO_2(g)$, $AgCl(s)$, $NH_3(g)$, $NH_3(aq)$, HCN, H_2O, $C_{12}H_{22}O_{11}$. Precipitates like AgCl are labeled with (s) to distinguish them from weak electrolytes in solution like NH_3 or HCN.[2]

We are now in a position to write ionic equations for the reactions in the preceding section.

Example 1   Mercury(II) nitrate and lithium chloride.

$$Hg^{++} + 2NO_3^- + 2Li^+ + 2Cl^- \rightarrow HgCl_2 + 2Li^+ + 2NO_3^-$$

All the salts but $HgCl_2$ are strong, soluble electrolytes and are represented by ionic formulas. Note that nitrate and lithium ions are unchanged. They are bystander ions and can be omitted to focus attention on the actual combination.

Essential reaction: $Hg^{++} + 2Cl^- \rightarrow HgCl_2$

Example 2   Ammonium chloride and sodium hydroxide.

$$NH_4^+ + Cl^- + Na^+ + OH^- \rightarrow NH_3(g) + H_2O + Na^+ + Cl^-$$

The salts ammonium chloride and sodium chloride are strong electrolytes, as is the strong base. Ammonia and water are weak electrolytes. No essential part is taken by sodium and chloride ions, so they may be omitted.

Essential reaction: $NH_4^+ + OH^- \rightarrow NH_3 + H_2O$

---

[2] Other methods of marking precipitates or gases in common use involve arrows or bars: $CO_2(g)$ or $CO_2\uparrow$ or $\overline{CO_2}$; AgCl(s) or AgCl$\downarrow$ or $\underline{AgCl}$. Whatever system is used must be used consistently.

Example 3   Hydrochloric acid and calcium acetate.

$$2H^+ + 2Cl^- + Ca^{++} + 2OAc^- \rightarrow Ca^{++} + 2Cl^- + 2HOAc$$

Hydrochloric acid and the two salts are strong, soluble electrolytes, but acetic acid[3] is a weak acid and is written in the molecular form. The bystander ions are calcium and chloride.

Essential reaction: $H^+ + OAc^- \rightarrow HOAc$

Example 4   Zinc oxide and nitric acid.

$$ZnO(s) + 2H^+ + 2NO_3^- \rightarrow Zn^{++} + 2NO_3^- + 2H_2O$$

The acid and salt are strong and ionic. Zinc oxide is only very slightly soluble in water, so it is represented by a molecular formula. As the equation shows, it dissolves in the acid; the essential reaction must show this as well as the acid–base reaction. Only nitrate ions are bystanders.

Essential reaction: $ZnO(s) + 2H^+ \rightarrow Zn^{++} + H_2O$

Example 5   Silver nitrate and nickel(II) chloride.

$$2Ag^+ + 2NO_3^- + Ni^{++} + 2Cl^- \rightarrow 2AgCl(s) + Ni^{++} + 2NO_3^-$$

Essential reaction: $Ag^+ + Cl^- \rightarrow AgCl(s)$

Example 6   Sulfur dioxide and barium hydroxide.

$$SO_2(g) + Ba^{++} + 2OH^- \rightarrow BaSO_3(s) + H_2O$$

Nothing is superfluous here.

Example 7   Aqueous ammonia and copper(II) nitrate.

$$4NH_3 + Cu^{++} + 2NO_3^- \rightarrow Cu(NH_3)_4^{++} + 2NO_3^-$$

Essential reaction: $4NH_3 + Cu^{++} \rightarrow Cu(NH_3)_4^{++}$

Example 8   Magnesium iodide and aqueous bromine.

$$Mg^{++} + 2I^- + Br_2 \rightarrow Mg^{++} + 2Br^- + I_2$$

Essential reaction: $2I^- + Br_2 \rightarrow 2Br^- + I_2$

We now examine in more detail some of the principal types of reaction.

1.5   Proton Transfer Reactions

To recognize proton transfer reactions we must have some familiarity with proton donors (Brønsted acids) and acceptors (bases). There are no restrictions

---

[3] The formula HOAc is a convenient abbreviation for CH_3COOH.

as to electric charge; acids may be cations (NH_4^+), neutral molecules (H_2S), or anions (HSO_4^- or $H_2PO_4^-$). For each acid there is a corresponding or *conjugate base*, defined by

$$acid \leftrightharpoons base + proton$$

Thus the base conjugate to NH_4^+ is NH_3, and the one conjugate to $H_2PO_4^-$ is HPO_4^{--}. Except for acidic hydrides, such as H_2S, most inorganic acids contain hydrogen covalently bonded to oxygen; that is, they are acidic hydroxides:

$$H_2SO_4 \quad \text{is equivalent to} \quad SO_2(OH)_2$$

Acetic acid, the most familiar organic acid, is likewise $CH_3CO(OH)$; the acidic character of hydrogen atoms covalently bonded to carbon as in CH_3— can usually be ignored. Hydrated metal ions can also act as acids:

$$Al(H_2O)_6^{3+} \leftrightharpoons Al(OH)(H_2O)_5^{++} + proton$$

The pull of the small, triply charged Al^{3+} ion on the oxygen end of the water molecule can be thought of as causing a loosening of the O—H covalent bond, making it possible for a proton to be lost.

Here are some examples of proton transfer reactions.

(1) Ionization of an acid:

$$HCl + H_2O \rightarrow H_3O^+ + Cl^-$$

We have written the proton in its simplest hydrated form, H_3O^+, to emphasize that it becomes attached to a solvent molecule or molecules, rather than remaining free. For simplicity it is convenient to use the symbol H^+ to stand for the hydrated proton—we usually do not show the hydration of other ions. The ionization would then be written $HCl \rightarrow H^+ + Cl^-$, with the understanding that all three symbols stand for hydrated species.

(2) Neutralization of acetic acid by sodium hydroxide:

$$HOAc + OH^- \leftrightharpoons OAc^- + H_2O$$

(3) Hydrolysis of trisodium phosphate:

$$PO_4^{3-} + H_2O \leftrightharpoons HPO_4^{--} + OH^-$$

(4) Dissolution of magnesium hydroxide by ammonium chloride:

$$Mg(OH)_2(s) + 2NH_4^+ \rightarrow 2NH_3(g) + 2H_2O + Mg^{++}$$

Can you identify the proton donors and acceptors?

It is sometimes useful to think of these reactions as involving a competition between two bases for the proton. In the neutralization reaction, for example, the

two bases are OH^- and OAc^-

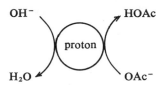

In this competition, the hydroxide ion usually wins, because it is a much stronger base than acetate—or most other common bases, for that matter. To express this concept quantitatively, we need to examine the equilibrium constants for these reactions.

1.6 The Extent of Proton Transfer Reactions

The strength of a base, such as OAc^-, relative to the solvent anion OH^- determines the equilibrium point of the reaction

$$OAc^- + H_2O \leftrightharpoons HOAc + OH^- \tag{1.1}$$

and is expressed by the equilibrium constant for this reaction

$$K_b(OAc^-) = \frac{[OH^-][HOAc]}{[OAc^-]} = 5.7 \times 10^{-10} M$$

It is common practice to refer to reaction (1.1) as hydrolysis and to K_b as the hydrolysis constant. Since reaction (1.1) is simply the standard reaction of the base OAc^- with solvent, we prefer to call K_b the *basicity constant*. The small value of this constant for acetate indicates a low degree of hydrolysis; acetate ion is a weak base.

For hydrolysis of the phosphate ion,

$$PO_4{}^{3-} + H_2O \leftrightharpoons HPO_4{}^{--} + OH^-$$

the basicity constant is $2.4 \times 10^{-2} M$. This is a much larger value than that for acetate. Phosphate ion is a stronger base than acetate ion. Hydrolysis of acetate is barely detectable; hydrolysis of phosphate is extensive, though still not complete.

Nothing that has been written so far restricts basicity constants to anion bases; K_b for ammonia is the equilibrium constant for its standard reaction with water:

$$NH_3 + H_2O \leftrightharpoons NH_4{}^+ + OH^- \qquad K_b(NH_3) = \frac{[NH_4{}^+][OH^-]}{[NH_3]}$$

$$= 1.8 \times 10^{-5} M$$

The strength of an acid, such as HOAc, relative to the hydrated proton determines the equilibrium point for the reaction represented by

$$HOAc + H_2O \leftrightharpoons H_3O^+ + OAc^-$$

or more simply by

$$HOAc \leftrightharpoons H^+ + OAc^-$$

The acid strength is expressed by an *acidity constant*:

$$K_a(HOAc) = \frac{[H^+][OAc^-]}{[HOAc]} = 1.8 \times 10^{-5} M$$

It will be noted that for a conjugate acid–base pair, such as HOAc–OAc$^-$, the product of acidity and basicity constants is the ion product of water

$$K_a \times K_b = \frac{[H^+]\cancel{[OAc^-]}}{\cancel{[HOAc]}} \times \frac{\cancel{[HOAc]}[OH^-]}{\cancel{[OAc^-]}} = K_w$$

Thus for acetic acid at 25°C,

$$K_a(HOAc) \times K_b(OAc^-) = 1.8 \times 10^{-5} \times 5.7 \times 10^{-10} = 1.0 \times 10^{-14} M^2$$

and for the phosphate pair at 25°C,

$$K_a(HPO_4^{--}) \times K_b(PO_4^{3-}) = 4.2 \times 10^{-13} \times 2.4 \times 10^{-2} = 1.0 \times 10^{-14} M^2$$

Given two of the three constants, say K_a and K_w, we can always find the third, K_b. Values of K_a and K_b for common acids and bases at 25°C and of the ion product of water at various temperatures are given in Appendixes A.2 and A.3, respectively.

Equilibrium constants for other proton transfer reactions can be obtained by combination of known constants. For neutralization of acetic acid, for example,

$$HOAc + OH^- \leftrightharpoons H_2O + OAc^-$$

the equilibrium constant expression is

$$K = \frac{[OAc^-]}{[HOAc][OH^-]} = \frac{1}{K_b(OAc^-)} = \frac{K_a(HOAc)}{K_w} = 1.8 \times 10^9$$

The large value of this constant indicates that this reaction, which is the reverse of hydrolysis of the acetate ion, goes almost to completion.

1.7 Precipitation Reactions

A distinction should be made at the outset between formation of a weak electrolyte, which involves making a covalent bond, and formation of a precipitate, which in most cases results from clustering of ions into an ionic lattice. These are different processes, and the solubility of a substance in water has no direct connection with whether it is a strong or weak electrolyte. To bring the point home, consider the two chlorides of mercury: $HgCl_2$ is a weak electrolyte and soluble in water, whereas Hg_2Cl_2 is a strong electrolyte and virtually insoluble

(Sec. 7.4). True, a saturated solution of the latter salt has a low conductivity relative to that of 0.1 M $CaCl_2$, but to be fair we should compare the two at the same molar concentration—then the difference is not so striking.

Attempts to predict the solubility of a substance usually start with the premise that dissolving a substance is equivalent to tearing a crystal apart into gaseous ions and then hydrating them and bringing them into solution. This is a perfectly sound point of view but one that has yet to produce useful results. The trouble with such theoretical calculations is that the solubility has to be obtained from the small difference between two large energy changes. We are in the position of trying to find the weight of an engineer by weighing the locomotive before and after he climbs into the cab.

The state of chemical theory being what it is, we have to fall back on empirical solubility rules for practical guidance. A brief summary is given in Table 1.2; a

Table 1.2

Summary of Solubilities in Water

Rule	Exceptions
1. Nitrates and acetates are generally soluble.	No common exceptions. Silver acetate is moderately insoluble.
2. Compounds of the alkali metals and ammonium ion are generally soluble.	No common exceptions. Some that are moderately insoluble are formed in Cation Group 5.
3. Fluorides are generally insoluble.	Fluorides of the alkali metals, ammonium ion, Ag(I), Al, Sn, and Hg.
4. Chlorides, bromides, and iodides are generally soluble.	The halides of Ag(I), Hg(I), and Pb(II); HgI_2; BiOCl and SbOCl.
5. Sulfates are generally soluble.	$PbSO_4$, $SrSO_4$, $BaSO_4$; $CaSO_4$ and Hg_2SO_4 are moderately insoluble.
6. Carbonates and sulfites are generally insoluble.	Those of the alkali metals and ammonium.
7. Sulfides are generally insoluble.	Those of the alkali metals and ammonium. Sulfides of the alkaline earth metals and Cr_2S_3 and Al_2S_3 are decomposed by water.
8. Hydroxides are generally insoluble.	Those of the alkali metals and ammonium. The hydroxides of Ba, Sr, and Ca are moderately soluble.

more complete set of rules is in Appendix A.1. The following examples show how we make predictions on the basis of the rules.

Example 1   Sodium sulfite and zinc nitrate. Of the possible products, sodium nitrate and zinc sulfite, the first should be a soluble, strong electrolyte, the second a precipitate. Both reactants are soluble, strong electrolytes, so that the essential reaction is

$$Zn^{++} + SO_3^{--} \rightarrow ZnSO_3(s)$$

Example 2   Magnesium sulfate and ammonium acetate. Both of the possible products, magnesium acetate and ammonium sulfate, are soluble, strong electrolytes. No reaction will occur in dilute solution.

1.8   The Solubility Product and Extent of Reaction

The solubility product is the constant for equilibrium between a precipitate of a slightly soluble substance and its ions in the saturated solution; for example,

$$PbI_2(s) \leftrightharpoons Pb^{++} + 2I^- \qquad K_{sp} = [Pb^{++}][I^-]^2 = 7.1 \times 10^{-9} M^3$$

When precipitation of lead iodide occurs by mixing potassium iodide and lead nitrate solutions, the chemical equation is the right-to-left portion of that above, and the equilibrium constant is the reciprocal of the solubility product:

$$Pb^{++} + 2I^- \rightarrow PbI_2(s) \qquad K = 1/K_{sp} = 1.4 \times 10^8 M^{-3}$$

The large value of this constant indicates that if reasonably large concentrations of the ions are used, precipitation is bound to occur to a large extent. Suppose, however, that we mix the ions at very low concentrations. Will any reaction occur? If both ions are present at 0.0060 M concentration, then

$$[Pb^{++}][I^-]^2 = (6.0 \times 10^{-3})^3 = 2.2 \times 10^{-7} M^3$$

This is larger than the value $7.1 \times 10^{-9} M^3$ allowed at equilibrium. The solution is supersaturated, and precipitation will occur until the ion concentrations have been reduced to the point where their product is equal to the solubility product.

The solubility product constant can be used to determine the direction of competitive reactions.

Example 1   Can lead chloride be converted to lead chromate by adding potassium chromate to a saturated solution of the former salt? The postulated reaction is

$$PbCl_2(s) + CrO_4^{--} \rightleftharpoons PbCrO_4(s) + 2Cl^-$$

for which the equilibrium constant is

$$K = \frac{[Cl^-]^2}{[CrO_4^{--}]} = \frac{[Pb^{++}][Cl^-]^2}{[Pb^{++}][CrO_4^{--}]}$$

$$= \frac{K_{sp}(PbCl_2)}{K_{sp}(PbCrO_4)} = \frac{1.6 \times 10^{-5}}{2 \times 10^{-16}} = 8 \times 10^{10} M$$

The large value of K indicates that reaction should be very extensive unless the chromate ion concentration is kept very low.

Example 2   Will hydrochloric acid dissolve barium sulfate? The equilibrium is

$$BaSO_4(s) + H^+ \leftrightharpoons Ba^{++} + HSO_4^-$$

This is a competition for sulfate between barium ion and hydrogen ion. The overall equilibrium constant is a combination of the solubility product of $BaSO_4$ and the acidity constant of HSO_4^-:

$$K = \frac{[Ba^{++}][HSO_4^-]}{[H^+]} = [Ba^{++}][SO_4^{--}]\frac{[HSO_4^-]}{[H^+][SO_4^{--}]}$$

$$= \frac{K_{sp}(BaSO_4)}{K_a(HSO_4^-)} = \frac{1.0 \times 10^{-10}}{1.0 \times 10^{-2}} = 1.0 \times 10^{-8}\ M$$

The small value of this constant indicates that very little barium sulfate will be dissolved.

When equilibrium constants are close to unity, the direction of reaction may be altered by changes in concentration.

Example 3   To what extent will 500 mg of barium sulfate be transposed to barium carbonate by 10 ml of 1.0 M Na_2CO_3? The equilibrium is

$$BaSO_4(s) + CO_3^{--} \leftrightharpoons BaCO_3(s) + SO_4^{--}$$

with a constant

$$K = \frac{[SO_4^{--}]}{[CO_3^{--}]} = \frac{K_{sp}(BaSO_4)}{K_{sp}(BaCO_3)} = 0.062$$

Let $x = [SO_4^{--}]$ at equilibrium; then $1.0 - x = [CO_3^{--}]$ at equilibrium.[4] The equilibrium constant can be expressed as $x/(1-x)$. Putting this equal to 0.062, we obtain $x = 0.059\ M$. The extent of conversion is the ratio of the millimoles of sulfate in solution to the millimoles of barium sulfate taken:

$$\% \text{ conversion} = \frac{10\ ml \times 0.059\ mmol\ ml^{-1}}{500\ mg/233\ mg\ mmol^{-1}} \times 100 = 28\%$$

To obtain a greater conversion, we would have to use either a larger volume of 1.0 M Na_2CO_3 or 10 ml of a more concentrated solution of this reagent.

1.9   Formation of Complex Ions

Complex ions are formed when neutral molecules like ammonia or anions like chloride bond to a central metal ion. The groups about the metal ion are called

---

[4] These stoichometric relationships are based on the assumption that sulfate and carbonate are involved only in the specified equilibrium. Then loss of a carbonate means gain of a sulfate, with the total concentration fixed at 1.0 M. The two ions also hydrolyze to HSO_4^- and HCO_3^-, but the error committed in neglecting these additional equilibria is small.

ligands, and the number of points of attachment to the metal ion is its *coordination number.* The most common coordination numbers are 4, as in $HgCl_4^{--}$, and 6, as in $Co(NO_2)_6^{3-}$, but silver(I) ion commonly has a coordination number of 2, as in $Ag(NH_3)_2^+$.

Complexes may be divided into two classes: *labile* and *inert.* The latter, which include $Co(NH_3)_6^{3+}$, $Cr(H_2O)_6^{3+}$, and $Fe(CN)_6^{4-}$, form or dissociate very slowly, so that they may not be in true equilibrium with their components. While their very inertness was useful in the development of Werner's theory of co-ordination compounds and in their employment as laboratory reagents, it means they are less interesting in discussions of rapid ionic reactions. The labile complexes dissociate and form rapidly and are frequently encountered in the laboratory work of qualitative analysis.

Since cations in aqueous solution are generally hydrated, complex formation occurs by stepwise substitution of ligands for water molecules in the coordination shell of the ion. The extent of reaction in aqueous solution therefore is determined by competition between water molecules and ligands for the metal ion. This must not be lost sight of even though we customarily omit water of hydration in writing ionic formulas:

$$Ag^+ + NH_3 \leftrightarrows AgNH_3^+ \qquad K_1 = \frac{[AgNH_3^+]}{[Ag^+][NH_3]} = 2.1 \times 10^3 \ M^{-1}$$

$$AgNH_3^+ + NH_3 \leftrightarrows Ag(NH_3)_2^+ \qquad K_2 = \frac{[Ag(NH_3)_2^+]}{[AgNH_3^+][NH_3]} = 7.7 \times 10^3 \ M^{-1}$$

The quotients K_1 and K_2 are referred to as stepwise formation constants of the complexes. Their product β_2 is the *overall formation constant*[5]

$$\beta_2 = K_1 \times K_2 = \frac{[Ag(NH_3)_2^+]}{[Ag^+][NH_3]^2} = 1.6 \times 10^7 \ M^{-2}$$

The reciprocal of β_2 is often tabulated and is called the instability constant of the complex. Its small value, $6.3 \times 10^{-8} \ M^2$, for $Ag(NH_3)_2^+$ indicates that this complex is weakly dissociated.

As an example of a reaction forming a complex, let us consider the dissolution of silver chloride in aqueous ammonia,

$$AgCl(s) + 2NH_3 \leftrightarrows Ag(NH_3)_2^+ + Cl^-$$

---

[5] The subscript 2 marks β as the constant for addition of two ligands to the metal ion. Similarly, the overall constant for addition of 4 NH_3 to Cu^{++} to give $Cu(NH_3)_4^{++}$ will be designated β_4. The overall equilibrium constant for addition of 2 H^+ to 2 CrO_4^{--} in the reaction represented by

$$2H^+ + 2CrO_4^{--} \leftrightarrows Cr_2O_7^{--} + H_2O$$

is designated β_{22} (Sec. 9.13). These conventions are those adopted in the authoritative compilation of stability constants: Special Publication 17 of The Chemical Society, London.

for which the equilibrium constant is

$$K = \frac{[Ag(NH_3)_2{}^+][Cl^-]}{[NH_3]^2} = \frac{[Ag(NH_3)_2{}^+]}{[Ag^+][NH_3]^2} \times [Ag^+][Cl^-]$$

$$= \beta_2 \times K_{sp}(AgCl) = 1.6 \times 10^7 \times 1.8 \times 10^{-10} = 2.9 \times 10^{-3}$$

The small value of this quotient indicates that the reaction occurs to only a slight extent when all three solute species are present at moderate to high concentrations. Silver chloride can actually be made to dissolve completely by increasing the concentration of ammonia.

1.10 Systematic Inorganic Chemistry. Reactions of Silver(I) Ion

We have seen how known equilibrium constants $(K_a, K_w, K_{sp}, \beta)$ can be combined to obtain constants for various ionic reactions. These constants provide the basis for arranging various reactions of an ion in a definite order. This is illustrated in Table 1.3 for Ag^+ ion. Such a tabulation enables us to predict other reactions. Any given silver salt or complex ought to be transformable into any one of those below it in the table. Silver chromate, for example, should be convertible to silver chloride:

$$Ag^+ + Cl^- \rightarrow AgCl(s) \qquad\qquad \log K = 9.74$$

$$-(Ag^+ + \tfrac{1}{2}CrO_4{}^{--} \rightarrow \tfrac{1}{2}Ag_2CrO_4(s) \qquad\qquad \log K = 5.81)$$

$$\overline{\tfrac{1}{2}Ag_2CrO_4(s) + Cl^- \rightarrow AgCl(s) + \tfrac{1}{2}CrO_4{}^{--} \qquad \log K = 3.93}$$

A large value of log K means a large equilibrium quotient or extensive reaction. Although Table 1.3 gives only 15 reactions of silver ion, 105 other combinations are possible, making 120 reactions in all.

The position of the formation reaction for $Ag(NH_3)_2{}^+$ ion would seem to indicate that AgCl would not dissolve in aqueous ammonia. As we saw in the preceding section, the extent of formation of a complex is sensitive to changes in the concentration of its ligand. At high concentrations of ammonia the position of the formation reaction for $Ag(NH_3)_2{}^+$ would be shifted below that for silver chloride.

1.11 Redox Reactions. Oxidation Numbers

In some reactions, simple ions suffer an increase or decrease of charge; for example,

$$Fe^{3+} + V^{++} \rightarrow V^{3+} + Fe^{++}$$

$$Cl_2 + 2Br^- \rightarrow Br_2 + 2Cl^-$$

Table 1.3

Selected Reactions of Silver (I) Ion

Typical reagents*	Equation	Color of product	log K at 25°C	$\Delta G°$
$(NH_4)_2SO_4$	$Ag^+ + \frac{1}{2}SO_4^{--} \leftrightarrows \frac{1}{2}Ag_2SO_4(s)$	white	2.38	−3.3
NaOAc	$Ag^+ + OAc^- \leftrightarrows AgOAc(s)$	white	2.64	−3.6
K_2CO_3, $NaHCO_3$, not $(NH_4)_2CO_3$	$Ag^+ + \frac{1}{2}CO_3^{--} \leftrightarrows \frac{1}{2}Ag_2CO_3(s)$	white	5.54	−7.6
K_2CrO_4	$Ag^+ + \frac{1}{2}CrO_4^{--} \leftrightarrows \frac{1}{2}Ag_2CrO_4(s)$	purplish red	5.81	−7.9
Na_2HAsO_4	$Ag^+ + \frac{1}{3}AsO_4^{3-} \leftrightarrows \frac{1}{3}Ag_3AsO_4(s)$	reddish brown	7.0	−9.2
$1\ M\ NH_3$	$Ag^+ + 2NH_3 \leftrightarrows Ag(NH_3)_2^+$	colorless	7.23	−9.9
NaOH, $Ba(OH)_2$	$Ag^+ + OH^- \leftrightarrows \frac{1}{2}Ag_2O(s) + \frac{1}{2}H_2O$	brownish black	7.71	−10.5
HCl, $NH_4Cl^†$	$Ag^+ + Cl^- \leftrightarrows AgCl(s)$	white	9.74	−13.3
NH_4SCN, $KSCN^†$	$Ag^+ + SCN^- \leftrightarrows AgSCN(s)$	white	12.0	−16.4
$KBr^†$	$Ag^+ + Br^- \rightleftharpoons AgBr(s)$	pale yellow	12.28	−16.7
$Na_2S_2O_3$, in excess	$Ag^+ + 2S_2O_3^{--} \leftrightarrows Ag(S_2O_3)_2^{3-}$	colorless	13.46	−18.4
$KCN^†$	$Ag^+ + CN^- \leftrightarrows AgCN(s)$	white	15.70	−21.4
$KI^†$	$Ag^+ + I^- \leftrightarrows AgI(s)$	yellow	16.08	−22.0
KCN, in excess	$Ag^+ + 2CN^- \leftrightarrows Ag(CN)_2^-$	colorless	19.85	−27.1
H_2S, Na_2S	$Ag^+ + \frac{1}{2}S^{--} \leftrightarrows \frac{1}{2}Ag_2S(s)$	black	24.6	−33.6

Source: *General Chemistry* by Martin A. Paul, Edward J. King, and Larkin H. Farinholt. New York: Harcourt Brace Jovanovich, 1967, p. 494.

* Common laboratory reagents are given. Other soluble reagents, such as the potassium, sodium, and ammonium salts could be used unless the ammonium compound furnishes sufficient ammonia to give the ammonia complex.

† These reagents redissolve the precipitates, if used in excess.

Let us define a set of *oxidation numbers* for each element such that

   (1) The oxidation number of an ion is equal to its charge.

   (2) The oxidation number of a free or uncombined element is zero.

An increase in oxidation number is then defined as *oxidation* and a decrease as *reduction*. In the examples just given, V^{++} and Br^- ions are oxidized and Cl_2 and Fe^{3+} ion are reduced. These changes are brought about by transfers of electrons:

oxidation by loss of electrons,

$$V^{++} \rightarrow V^{3+} + e^-$$

$$2Br^- \rightarrow Br_2 + 2e^-$$

and reduction by gain of electrons,

$$e^- + Fe^{3+} \rightarrow Fe^{++}$$

$$2e^- + Cl_2 \rightarrow 2Cl^-$$

These concepts can be generalized to include molecules and ions containing covalent bonds. Oxidation numbers have no simple physical meaning with these more complex particles. They are assigned according to certain arbitrary rules, which include the two already given and the following three:

(3) The oxidation number of combined oxygen is usually $2-$. It is $1-$ in peroxides and $\frac{1}{2}-$ in superoxides.

(4) The oxidation number of combined hydrogen is usually $1+$. In the hydrides of the active metals—for example, LiH and CaH_2—it is $1-$.

(5) The sum of the oxidation numbers of the elements in an ion must be equal to the net ionic charge, and the sum for a molecule must be zero.

These rules are illustrated by the following examples.

Example 1 Find the oxidation number of carbon in acetic acid, CH_3COOH. The total oxidation number of 4 H is $4+$ and that of 2 O is $4-$. If x is the oxidation number of each C, we have $2x + 4 - 4 = 0$ or $x = 0$. This is not a unique assignment: we could call one C $4+$ and the other $4-$, or we could assign different oxidation numbers to oxygen and hydrogen. The only requirement of any physical significance is that the sum must be zero for the molecule.

Example 2 Find the oxidation number of chromium in the dichromate ion, $Cr_2O_7^{--}$. If x is the oxidation number of each Cr, we have $2x - 14 = -2$, or $x = 6+$.

The oxidation number concept is useful in organizing the chemistry of elements that form several ions. Consider the oxidation states of chromium shown in Fig. 1.1. The principal positive oxidation states of chromium are $2+$, $3+$, and $6+$; the illustration shows the various manifestations of these states in acidic and basic solutions.

By contrast with chromium, the oxidation numbers of some other elements are virtually invariant. For example, the alkali metals (Group IA of the periodic table) form only simple univalent ions, so you can assume that in their compounds they always have an oxidation number of $1+$. The elements of Group IIA can always be assumed to have oxidation numbers of $2+$, and the halogens, except in their compounds with oxygen, almost always have oxidation numbers of $1-$.

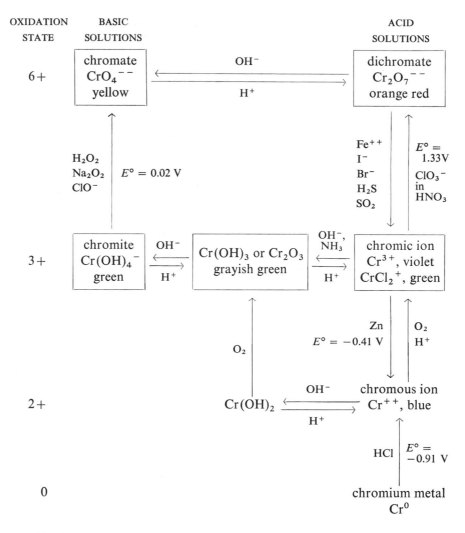

Fig. 1.1. The oxidation states of chromium. The important compounds and ions are shown in rectangles.

1.12 Balancing Equations for Redox Reactions

Various recipes may be used to obtain a balanced equation for a redox reaction with a minimum of effort. None of these give any real indication of the sequence of events that actually take place; that is usually far too complex to be shown by a single equation. A procedure that establishes the correct proportions of reactants and products is still very useful.

The following procedure, often called the *ion–electron method*, is particularly convenient when several elements are being oxidized or reduced in one reaction. There are six steps in the procedure.

(1) Write skeleton partial equations showing separately the oxidation and the reduction.

Example The oxidation of phosphorous acid by potassium dichromate in acid solution. The skeleton equations are:

$$Cr_2O_7^{--} \rightarrow Cr^{3+}$$

$$H_3PO_3 \rightarrow H_3PO_4$$

The products must be given or predicted on the basis of experience (see Fig. 1.1).

(2) Equalize the number of atoms on both sides of each partial equation. It is generally best to proceed in the following order.

(*a*) First balance the atoms of the elements undergoing changes in oxidation number.

Example

$$Cr_2O_7^{--} \rightarrow 2Cr^{3+}$$

(*b*) Next balance the oxygen atoms. For this purpose we use common currency: H_2O, H^+, and OH^-, always present in aqueous solutions, rather than exotic species like atomic O or O^{--} ions. The following rules can then be used.

Rule A: $2H^+ + (\text{extra O}) \leftrightharpoons H_2O$ (acid solution)

Rule B: $H_2O + (\text{extra O}) \leftrightharpoons 2OH^-$ (basic solution)

Note that when read from left to right, the rules show how to dispose of an extra oxygen, and from right to left, how to furnish one.

Example In $Cr_2O_7^{--} \rightarrow 2Cr^{3+}$ there are seven extra oxygen atoms on the left. Therefore, balance the equation by adding 14 H^+ to the left and 7 H_2O to the right as specified in Rule A:

$$14H^+ + Cr_2O_7^{--} \rightarrow 2Cr^{3+} + 7H_2O$$

In $H_3PO_3 \rightarrow H_3PO_4$ there is an extra oxygen atom on the right; add H_2O to the left and 2 H^+ to the right:

$$H_2O + H_3PO_3 \rightarrow H_3PO_4 + 2H^+$$

(*c*) If extra hydrogen atoms are still to be disposed of, use Rules C and D.

Rule C: $(\text{extra H}) \leftrightharpoons H^+$ (acid solution)

Rule D: $(\text{extra H}) + OH^- \leftrightharpoons H_2O$ (basic solution)

(3) Balance each partial equation electrically by adding electrons to equalize *ionic* charges.

Examples

$$14H^+ + Cr_2O_7^{--} + 6e^- \rightarrow 2Cr^{3+} + 7H_2O$$

$$+14 \qquad -2 \qquad -6 = +6 \quad + \quad 0$$

$$H_2O + H_3PO_3 \rightarrow H_3PO_4 + 2H^+ + 2e^-$$

$$0 \; + \; 0 \; = \; 0 \; + \; 2 \; - \; 2$$

The electron changes are found without reference to the oxidation numbers; only the ionic charges are used. The same result would be obtained using oxidation numbers: 2 Cr(VI) to 2 Cr(III) is a total change of 6 balanced by the gain of 6 electrons.

(4) By cross multiplication, make the number of electrons lost in one partial equation equal to the number gained in the other.

Example

$$(H_2O + H_3PO_3 \rightarrow H_3PO_4 + 2H^+ + 2e^-) \times 3$$

(5) Add the two partial equations and cancel electrons, H^+, H_2O, and OH^- as far as possible. There should be no electrons in the final equation.

Example

$$8H^+ + 3H_3PO_3 + Cr_2O_7^{--} \rightarrow 3H_3PO_4 + 2Cr^{3+} + 4H_2O$$

(6) Check the final equation.
(*a*) The net ionic charges on the two sides should be equal.

Example   Left side 0, right side 0.

(*b*) The number of atoms of each kind must be the same on both sides.

Example   Each side has 17 H, 16 O, 3 P, and 2 Cr.

For reactions in basic solutions, as an alternative to balancing oxygen atoms by the rule given above, you may determine the number of electrons to be added from the change in oxidation number and then balance charges with hydroxide ions. Consider, for example, reduction of hypochlorite in basic solution: ClO^- to Cl^-. The oxidation number of chlorine changes from $1+$ to $1-$, so that two electrons must be gained:

$$2e^- + ClO^- \rightarrow Cl^-$$

The negative charges are balanced if we add 2 OH⁻ to the right; we then have to add H_2O to the left to balance hydrogen atoms:

$$2e^- + ClO^- + H_2O \rightarrow Cl^- + 2OH^-$$

The procedure of using oxidation number changes to determine the number of electrons required has the advantage of focusing attention on the atoms that are oxidized and reduced. On the other hand, it can be cumbersome in balancing changes such as,

$$CH_3CH_2OH \rightarrow CH_3CHO$$

or confusing when more than one atom is oxidized or reduced, as in

$$As_2S_3(s) \rightarrow S(s) + H_2AsO_4^-$$

Try balancing these partial equations—for reaction in acid solution—by both the ion–electron and the oxidation number methods.

1.13   The Extent of Redox Reactions

We could formulate equilibrium constants for redox reactions, as we have done for other types of reactions. Because galvanic cells are convenient devices for studying such reactions, the half-reactions of the ion–electron method being the electrode reactions in a cell, we customarily express the extent of redox reactions by a potential difference or *electromotive force E*. When all substances involved in the reaction are in their standard states, which for solutes in solution means at approximately 1 M concentration, the electromotive force has a standard value designated $E°$. The equilibrium constant for the cell reaction and $E°$ are related by

$$E° = (k_N/n)\log K \qquad (1.2)$$

The Nernst factor k_N has the value 0.0592 V at 25°C, and n is the number of electrons transferred in the chemical equation on which the expression for K is based.

Values of $E°$ for complete reactions are obtained by combining standard electrode potentials for the half-reactions. The prevailing convention is to write all such half-reactions to correspond to reduction and to subtract one from the other to get the complete equation.

Example   Oxidation of sulfur dioxide by dichromate.

$$14H^+ + Cr_2O_7^{--} + 6e^- \rightarrow 2Cr^{3+} + 7H_2O \qquad\qquad E° = 1.33 \text{ V}$$

$$-3[4H^+ + SO_4^{--} + 2e^- \rightarrow SO_2 + 2H_2O] \qquad\qquad -(E° = 0.17 \text{ V})$$

$$\overline{2H^+ + Cr_2O_7^{--} + 3SO_2 \rightarrow 2Cr^{3+} + 3SO_4^{--} + H_2O \qquad\qquad E° = 1.16 \text{ V}}$$

The $E°$ values combine in the same way as the half-reactions except that multiplication by 3 is not required, since $E°$ values represent work per unit charge. A large positive value for $E°$, such as was obtained in this example, indicates a large extent of reaction when all solute species are at 1 M concentration. When $E°$ is near zero, changes in concentration may easily alter the direction of chemical reaction.

1.14 Gibbs Energy Changes and the Extent of Reaction

A third way of obtaining information about the extent of reaction is based on calorimetric data. From these are derived so-called free energies or *Gibbs energies* G. Extensive tables of standard Gibbs energies of formation of various species from their elements in their normal condition at 25°C and 1 atm are available.

Example

$$2AgCl(s) + S^{--} \rightarrow Ag_2S(s) + 2Cl^-$$

From the tables we find the following values:

(a) $2Ag(s) + S(rhombic) \rightarrow Ag_2S(s)$ $\Delta G_f° = -9.56 \text{ kcal mol}^{-1}$

(b) $Cl_2(g) \rightarrow 2Cl^-(aq)$ $\Delta G_f° = 2 \times (-31.35)$

(c) $2Ag(s) + Cl_2(g) \rightarrow 2AgCl(s)$ $\Delta G_f° = 2 \times (-26.22)$

(d) $S(rhombic) \rightarrow S^{--}(aq)$ $\Delta G_f° = 20.6$

The desired overall reaction is seen to be the combination $(a)+(b)-(c)-(d)$. The $\Delta G_f°$ values are combined in the same way

$$\Delta G° = -9.56 - 62.70 - (-52.44) - 20.6 = -40.5 \text{ kcal}$$

A large negative value like this one indicates extensive reaction when the ions are at unit concentration.

Any information about the extent of reaction supplied by K or $E°$ can be expressed equally well by $\Delta G°$ in view of the relationships

$$\Delta G° = -2.303RT \log K = -nFE° \tag{1.3}$$

where R is the ideal gas constant, T is the absolute temperature, and F is the Faraday of electric charge, 96,487 coulombs. The function of thermodynamics is to provide such relationships between seemingly unconnected properties. The three different ways of getting information about the extent of reaction, though connected by equation (1.3), persist because for a given reaction one type of measurement may be more convenient than another.

EXERCISES

1.1.   The conductivities of 0.0100 M solutions of NH_3 and NH_4Cl are, respectively, 1.23×10^{-4} and 14.13×10^{-4} ohm^{-1} $cm.^{-1}$. Why is the second so much larger than the first?

1.2.   Classify as strong or weak electrolytes: nitric acid, iron(II) sulfate, carbonic acid, barium hydroxide, mercury(II) chloride, arsenious acid, silver(I) nitrate, methylamine (a base), hydrofluoric acid, and ammonium acetate.

1.3.   The reaction of barium hydroxide and sulfuric acid is represented by the molecular equation

$$Ba(OH)_2 + H_2SO_4 \rightarrow BaSO_4 + 2H_2O$$

Rewrite it according to the conventions used for ionic equations. How could the progress of this reaction be followed by conductivity measurements? By a visual method?

1.4.   The following molecular equations represent reactions that actually take place. For each, decide what type of reaction is involved and write an ionic equation for the essential reaction.

(a) $Ca(NO_3)_2 + Na_2CO_3 \rightarrow CaCO_3 + 2NaNO_3$

(b) $CO_2 + 2KOH \rightarrow K_2CO_3 + H_2O$

(c) $AgI + 2Na_2S_2O_3 \rightarrow Na_3Ag(S_2O_3)_2 + NaI$

(d) $HNO_2 + NaHCO_3 \rightarrow NaNO_2 + H_2O + CO_2$

(e) $Mg(OH)_2 + (NH_4)_2SO_4 \rightarrow MgSO_4 + 2NH_3 + 2H_2O$

1.5.   Classify as proton donors and acceptors. (Some may be both; they are called ampholytes.) H_3O^+, OH^-, HNO_3, $CH_3NH_3^+$, $H_2PO_4^-$, HSO_4^-, H_2O, $Fe(H_2O)_6^{3+}$, S^{--}, $AgNH_3^+$

1.6.   Classify as soluble or virtually insoluble in water: PbI_2, $LiOH$, $Ca(OAc)_2$, $PbSO_4$, $(NH_4)_2CO_3$, KNO_3, NiS, $BiOCl$, $Mg(OH)_2$, Na_2SO_3, CaF_2, $(NH_4)_2S$, $CoSO_4$, AgI, $NaOH$

1.7.   Which of the following pairs of substances will combine? What type of reaction, if any, will occur? Write the ionic equation for the essential reaction.

(a) Nitrous acid and potassium hydroxide

(b) Iron(III) chloride and sodium hydroxide

(c) Nickel(II) chloride and aqueous ammonia

(d) Acetic acid and sodium chloride

(e) Zinc chloride and magnesium sulfate

1.8.   Give the oxidation number of the element specified for each substance.

(a) Manganese in $KMnO_4$	(b) Potassium in KCl
(c) Carbon in CO_2	(d) Chlorine in Cl_2
(e) Cobalt in $Co(NH_3)_6Cl_3$	(f) Chromium in K_2CrO_4
(g) Iron in $Fe_2(SO_4)_3$	(h) Aluminum in Na_3AlF_6
(i) Sulfur in $Na_2S_2O_4$	(j) Arsenic in H_3AsO_4

1.9. Separate the following skeleton equations into half-reactions, balance, and recombine them.

(a) $H_2S + Fe^{3+} \rightarrow Fe^{++} + S(s)$ (acid solution)

(b) $NO_3^- + Bi_2S_3(s) \rightarrow Bi^{3+} + NO(g) + S(s)$ (acid solution)

(c) $Fe_3O_4(s) + H_2O_2 \rightarrow Fe^{3+} + H_2O$ (acid solution)

(d) $HClO + Br_2 \rightarrow BrO_3^- + Cl^-$ (acid solution)

(e) $As_2S_3(s) + ClO_3^- \rightarrow Cl^- + H_2AsO_4^- + S(s)$ (acid solution)

(f) $Bi(OH)_3(s) + Sn(OH)_3^- \rightarrow Sn(OH)_6^{--} + Bi(s)$ (basic solution)

(g) $Al + NO_3^- \rightarrow Al(OH)_4^- + NH_3(g)$ (basic solution)

(h) $MnO_4^- + NO_2^- \rightarrow MnO_2(s) + NO_3^-$ (basic solution)

(i) $HO_2^- + Cr(OH)_4^- \rightarrow CrO_4^{--} + OH^-$ (basic solution)

(j) $Cu(NH_3)_4^{++} + S_2O_4^{--} \rightarrow SO_3^{--} + Cu(s) + NH_3$ (basic solution)

1.10. Consider the possible reaction of solid lead sulfate with a solution of sodium sulfide. Write an ionic equation for this reaction. From the solubility products of lead sulfate and lead sulfide (Appendix A.2), calculate the equilibrium constant for the reaction. Will the reaction go?

1.11. Consider the possible reaction represented by

$$2HOAc + Mg(OH)_2(s) \rightarrow Mg^{++} + 2OAc^- + 2H_2O$$

Formulate the expression for the equilibrium constant and show that it is a combination of K_a for acetic acid, K_{sp} for magnesium hydroxide, and K_w for water. Calculate the equilibrium constant from the values of these constants given in Appendix A. Will the reaction occur, assuming a concentration of acetic acid near 1 M?

1.12. Write an ionic equation for the transposition of strontium sulfate to strontium carbonate. Calculate the equilibrium constant from the solubility products given in Appendix A.2. Calculate the extent of conversion if 500 mg of strontium sulfate is treated with 10 ml of 1.0 M Na_2CO_3, and compare with the result obtained for barium sulfate in Sec. 1.8.

1.13. The reaction of copper(II) sulfide with acid is represented by

$$CuS(s) + 2H^+ \rightarrow Cu^{++} + H_2S$$

Combine the solubility product of the sulfide from Appendix A.2 with the acidity constants of H_2S and HS^- from Appendix A.3 to get the equilibrium constant for the reaction. Would you expect it to occur to an appreciable extent? How might the solubility of the sulfide in acid be increased?

1.14. Given the following standard electrode potentials

$$AgI(s) + e^- \rightarrow Ag(s) + I^- \qquad E^\circ = -0.152 \text{ V}$$

$$AgCl(s) + e^- \rightarrow Ag(s) + Cl^- \qquad E^\circ = +0.222 \text{ V}$$

$$2H^+ + 2e^- \rightarrow H_2(s) \qquad E^\circ = 0.000 \text{ V}$$

Will silver liberate hydrogen from 1 M hydroiodic acid? From 1 M hydrochloric acid?

1.15. Given the following standard electrode potentials

$$Cr^{3+} + e^- \rightarrow Cr^{++} \qquad E° = -0.41 \text{ V}$$
$$O_2(g) + 4H^+ + 4e^- \rightarrow 2H_2O \qquad E° = 1.229 \text{ V}$$

Will Cr^{++} ion be stable in an acid solution exposed to air? Calculate $E°$ for the overall change.

1.16. Calculate $\Delta G°$ for $PbS(s) + 2H^+ \rightarrow Pb^{++} + H_2S(g)$ from the following standard Gibbs energies of formation of the substances from their elements: $PbS(s)$, -22.15; H^+, 0; Pb^{++}, -5.81: $H_2S(g)$, -7.89 kcal mol^{-1}. What is the significance of the result?

T W O

Ionic Separations

2.1 Introduction

In analyzing complex mixtures it is often necessary to isolate one ion from the rest before it can be identified. This is easy to accomplish when silver ion, for example, forms an insoluble chloride and calcium ion does not. Simple ionic reactions like the precipitation of silver chloride will effect a separation when there are gross differences in solubility. While there are only a few insoluble chlorides, there are many insoluble sulfides, hydroxides, and carbonates. The precipitation of these substances can still be used for separating ions if we capitalize on subtle differences in solubility. Some control over the concentration of the precipitating ion—sulfide, hydroxide, or carbonate—must then be exercised if separation is to be successful.

2.2 Fractional Precipitation of Sulfates

To illustrate the problems involved in separating two ions, let us consider first a mixture containing barium and calcium salts. Both cations form insoluble sulfates, but there is a considerable difference in their solubility products:

$$K_{sp} = [Ba^{++}][SO_4^{--}] = 1.0 \times 10^{-10} \ M^2$$

$$K_{sp} = [Ca^{++}][SO_4^{--}] = 2.4 \times 10^{-5} \ M^2$$

27

If we increase the sulfate ion concentration of a solution containing the two cations by adding powdered sodium sulfate a little at a time, the less soluble barium sulfate should precipitate first. We shall assume thorough mixing during and after each addition of the salt and shall neglect any change in volume of the solution due to the additions.

The precise efficiency of the separation depends on the initial concentrations of the two cations: we suppose 0.1 mg of Ba^{++} and 5.0 mg of Ca^{++} per milliliter. For use with the solubility products, the concentrations must be in millimoles per milliliter (molarity units):

$$[Ba^{++}] = \frac{0.10 \text{ mg/ml}}{137 \text{ mg/mmol}} = 7.3 \times 10^{-4} \, M$$

$$[Ca^{++}] = \frac{5.0 \text{ mg/ml}}{40 \text{ mg/mmol}} = 1.2 \times 10^{-1} \, M$$

The conditions for precipitation of these sulfates are

Ba: $7.3 \times 10^{-4}[SO_4^{--}] > 1.0 \times 10^{-10}$ or $[SO_4^{--}] > 1.4 \times 10^{-7}$

Ca: $1.2 \times 10^{-1}[SO_4^{--}] > 2.4 \times 10^{-5}$ or $[SO_4^{--}] > 2.0 \times 10^{-4}$

Because barium sulfate requires a much lower concentration of sulfate, it should precipitate first.

Suppose that addition of sodium sulfate is continued until calcium sulfate also begins to precipitate. When the solution is simultaneously saturated with respect to both salts, the two solubility product expressions must be satisfied by the same concentration of sulfate ion, for there can be no distinction between sulfate ions that are in equilibrium with barium sulfate and those that are in equilibrium with calcium sulfate. Hence the ratio of the solubility products gives the ratio of the metal ion concentrations in the solution that is saturated with respect to both precipitates:

$$\frac{[Ca^{++}]}{[Ba^{++}]} = \frac{K_{sp}(CaSO_4)}{K_{sp}(BaSO_4)} = \frac{2.4 \times 10^{-5}}{1.0 \times 10^{-10}} = 2.4 \times 10^5$$

This is much larger than the initial molar ratio of 160 to 1 because most of the barium has been precipitated. When precipitation of the calcium sulfate is about to begin, all the calcium ions originally present are still in solution and $[Ca^{++}]$ is $1.2 \times 10^{-1} \, M$, the initial concentration. The concentration of barium left in solution at this point is given by

$$2.4 \times 10^5 = \frac{1.2 \times 10^{-1}}{[Ba^{++}]} \quad \text{or} \quad [Ba^{++}] = 5 \times 10^{-7} \, M$$

This is $100(5 \times 10^{-7})/(7.3 \times 10^{-4}) = 0.07\%$ of the barium originally present, so that $100.00 - 0.07 = 99.93\%$ has precipitated before calcium sulfate starts to form.

These calculations show how, in principle, it is possible to separate two ions by fractional precipitation, but they are necessarily idealized. A high degree of supersaturation is required for the formation of both precipitates. It would therefore be necessary to add more sodium sulfate than we have calculated. The barium sulfate formed in the first step will be far from pure, for as the experiment is carried out the crystals form rapidly and tend to incorporate impurities by adsorption or trapping, a process called *coprecipitation*. As the solid sodium sulfate dissolves, a high concentration of sulfate ion will build up around the solid particles. Since the original solution contained a large excess of calcium over barium, it is probable that, in the presence of a local excess of sulfate, some calcium sulfate will precipitate along with the barium compound. A better procedure for the precipitation is described in the next section.

For qualitative analysis we do not always require that precipitates be free of contamination, and fractional precipitation is a useful method of separating them. The concentration ratio of the ions should be very large when precipitation of the more soluble compound starts. A ratio greater than 10^5 is desirable; thus barium and strontium cannot be satisfactorily separated by fractional precipitation of the sulfates because the ratio $[Sr^{++}]/[Ba^{++}]$ is only 7.6×10^3. The concentration ratio for precipitation of the chromates of these two metal ions is more satisfactory, namely, 4×10^4, and the effectiveness of the separation is increased by control of the chromate ion concentration.

2.3 Precipitation from Homogeneous Medium

Instead of adding the precipitating reagent to the solution in small increments a tedious and not very discriminating technique, we can generate the reagent within the solution by a slow chemical reaction. This elegant method is called *precipitation from homogeneous medium*. Sulfate, for example, can be generated by slow hydrolysis of sulfamic acid, a reaction catalyzed by hydrogen ion

$$NH_2SO_3^- + H_2O \xrightarrow{H^+} NH_4^+ + SO_4^{--}$$

The rate law for the reaction is found to be approximately represented by

$$\text{Rate of formation of sulfate} = k\,[H^+][NH_2SO_3^-] \qquad (2.1)$$

The rate is a million times faster in 0.1 M HCl ($[H^+] = 0.1\ M$) than in pure water ($[H^+] = 10^{-7}\ M$). The rate constant k increases with rise in temperature, and the formation of sulfate, very slow at room temperature, occurs at a convenient speed near the boiling point. The separation of barium from calcium or even from strontium is neatly accomplished by using sulfamic acid, rather than sodium sulfate, as the source of sulfate ion.

Aside from its convenience, precipitation from a homogeneous medium has the great advantage of forming purer precipitates. Ordinary precipitation occurs when two solutions are mixed; the degree of supersaturation as precipitation begins is very high, and precipitation occurs rapidly. During the few seconds of mixing, the medium is not homogeneous, and local regions of it may momentarily contain very high concentrations of a reagent. Coprecipitation is then very likely to occur. If, instead, one reagent is generated slowly within the solution, the degree of supersaturation is always low and the precipitate forms slowly. Coprecipitation is not eliminated by slow growth, but it is much reduced.

2.4   Precipitation of Sulfides by Hydrogen Sulfide. Control of the Sulfide Ion Concentration

Sulfides are so important in qualitative analysis that a detailed discussion of their precipitation is required. Geochemists are also interested in the conditions of sulfide precipitation because of the prevalence of sulfide ores. The extent of precipitation under various conditions is shown in Table 2.1. The order of precipitation of the sulfides agrees in general with what we would expect from equilibrium principles.

In an aqueous solution saturated with hydrogen sulfide gas, there are three equilibria to consider:

$$H_2S(g) \leftrightharpoons H_2S(aq) \qquad K_p = \frac{[H_2S]}{P_{H_2S}}$$

$$H_2S(aq) \leftrightharpoons H^+ + HS^- \qquad K_1 = \frac{[H^+][HS^-]}{[H_2S]}$$

$$HS^- \leftrightharpoons H^+ + S^{--} \qquad K_2 = \frac{[H^+][S^{--}]}{[HS^-]}$$

The product of the constants is

$$K_p K_1 K_2 = \frac{[H^+]^2[S^{--}]}{P_{H_2S}}$$

or for a hydrogen sulfide pressure of 1 atm

$$K_{ip} = [H^+]^2[S^{--}] \qquad (2.2)$$

The ion product constant K_{ip} therefore holds for a solution saturated with hydrogen sulfide at 1 atm pressure. We see from this expression that in strongly acid solutions the sulfide ion concentration is low, and in basic solutions for which $[H^+]$ is very small, $[S^{--}]$ is high. The sulfide ion concentration is thus controlled by regulating the hydrogen ion concentration. The solubility products of some of the sulfides are given in Table 2.1. Although the accuracy of some of these values is not very high, they do show that the sulfides of Cation Group 2 have lower solubility products than those of Cation Group 3. The dividing line comes between cadmium sulfide, the most soluble member of Group 2, and zinc sulfide, the least soluble one of Group 3. It is found experimentally that if the sulfide ion

Table 2.1

The Extent of Precipitation of Sulfides in Various Media

	K_{sp}	In 6 M HCl	In 3 M HCl	In 1 M HCl	In 0.5 M HCl	In 0.2 M HCl	In buffer $[H^+]=10^{-2*}$	In buffer $[H^+]=10^{-6\dagger}$	In NH_3 + NH_4Cl + $(NH_4)_2S$
Cation Group 2									
As_2S_5		P	S	S	S	S	S	S	—‡
As_2S_3		P	P	P	P	P	P	P	—‡
HgS	3×10^{-52}	P	P	P	P	P	P	P	P
CuS	8×10^{-36}	P	P	P	P	P	P	P	P
Sb_2S_5		I	P	P	P	P	P	P	—‡
Sb_2S_3		—	P	P	P	P	P	P	—‡
SnS_2		—	I	P	P	P	P	P	—‡
SnS		—	—	P	P	P	P	P	I‡
Bi_2S_3	1×10^{-96}	—	—	P	P	P	P	P	P
PbS	8×10^{-28}	—	I	I	I	P	P	P	P
CdS	7×10^{-27}	—	—	I	P	P	P	P	P
Cation Group 3									
ZnS	8×10^{-25}	—	—	—	—	CP	S	P	P
CoS	8×10^{-23}	—	—	—	—	CP	CP	P	P
NiS	2×10^{-21}	—	—	—	—	CP	CP	P	P
FeS	5×10^{-18}	—	—	—	—	—	—	CP	P
MnS	1×10^{-11}	—	—	—	—	—	—	—	P

Source: Reprinted with permission from G. E. F. Lundell and J. I. Hoffman, *Outlines of Methods of Chemical Analysis*, 1938, John Wiley & Sons, Inc., New York, pp. 49–54, and from Ernest H. Swift, *Introductory Quantitative Analysis*, 1950, W. H. Freeman and Co., San Francisco, California, pp. 422–423.

Key: *P*, complete or almost complete precipitation; *I*, incomplete precipitation; *S*, slow but eventually complete precipitation; *CP*, not precipitated alone but coprecipitated with other sulfides to an extent dependent on the other sulfides present; —, no precipitate.

* Buffers that maintain $[H^+]=10^{-2}$ M can be prepared from mixtures of sulfate and hydrogen sulfate ions or of formic acid and a formate.

† Buffers that maintain $[H^+]=10^{-6}$ M can be prepared from mixtures of acetic acid and an acetate.

‡ These sulfides do not precipitate in strongly basic solutions because of the formation of thiocomplex ions such as AsS_4^{3-}, SbS_3^{3-}, and SnS_3^{--}.

is controlled by adjustment of the concentration of hydrochloric acid to between 0.1 and 0.3 M, zinc sulfide can be kept in solution while cadmium sulfide (and the other sulfides of Group 2) is precipitated:

$$K_{sp}(CdS) < [Cd^{++}][S^{--}] \qquad K_{sp}(ZnS) > [Zn^{++}][S^{--}]$$

Some zinc is always coprecipitated with the sulfides of Group 2, even though zinc sulfide cannot be precipitated from a solution of a pure zinc salt in this acidic solution.

2.5   Precipitation of Sulfides by Thioacetamide

Thioacetamide is used to precipitate sulfides partly because of the unpleasant and dangerous character of hydrogen sulfide and partly because of the advantages of precipitation from homogeneous medium. Sulfides obtained by hydrolysis of thioacetamide are coarser, more crystalline precipitates than those obtained with hydrogen sulfide. They are easier to settle by centrifugation and are less contaminated by coprecipitated ions.

The hydrolysis of thioacetamide produces primarily acetamide[1] and hydrogen sulfide:

$$CH_3C(=S)NH_2 + H_2O \rightarrow CH_3C(=O)NH_2 + H_2S$$

This reaction, like the hydrolysis of sulfamic acid, is catalyzed by acid. In dilute acid solutions the rate law is

$$\text{Rate of formation of } H_2S = k\,[H^+][CH_3CSNH_2] \qquad (2.3)$$

The rate constant k is 0.019 liter mol^{-1} min^{-1} at 60° and 0.21 at 90°. Hence

$$\frac{\text{rate at } 90°}{\text{rate at } 60°} = \frac{0.21}{0.019} = 11$$

if the concentrations of hydrogen ion and thioacetamide are the same at the two temperatures. This means that if it takes an hour to produce a certain concentration of hydrogen sulfide at 60°, it takes only a little more than 5 minutes to achieve the same result at 90°. Clearly, it is advantageous to carry out precipitations of sulfides with thioacetamide as close to the boiling point as possible. The rate law also shows that a rapid reaction is favored by a high concentration of thioacetamide, so the reagent is added as a concentrated (8–13%) solution.

Experience has shown that most sulfides are precipitated at about the same acidities by hydrogen sulfide and thioacetamide. The principal exceptions are cadmium and arsenic(V) sulfides. Precipitation of cadmium sulfide by thioacetamide is frequently incomplete unless the acid concentration is reduced to 0.1 M.

---

[1] Thioacetamide differs from acetamide in having sulfur in place of oxygen. The prefix *thio-* is derived from the Greek word for sulfur.

A quantitative separation of cadmium from zinc is still possible, for the precipitation of zinc sulfide does not occur until the hydrogen ion concentration is reduced to about 0.01 M. Arsenic(V) is reduced by thioacetamide and precipitation of As_2S_3 is rapid, whereas the action of hydrogen sulfide on arsenic(V) is slow and troublesome.

In acid solutions with pH less than 3, the rate of precipitation of sulfides is governed by the rate of hydrolysis of thioacetamide [equation (2.3)].[2] It follows that if the acidity of the solution is increased, the precipitation will occur more rapidly. This will not continue indefinitely, however, because the increase in hydrogen ion concentration also lowers the sulfide ion concentration as required by the ion product [equation (2.2)]. Precipitation stops when, for example, $[Zn^{++}][S^{--}]$ falls below K_{sp}, the equilibrium value.

In basic solution, hydrolysis of thioacetamide follows a more complex course, and the rate of precipitation is much faster than it is in acid solution.

2.6   Numerical Calculations for the Precipitation of Sulfides

To test the equilibrium principles discussed in Secs. 2.4 and 2.5, let us try a numerical calculation. At 100°C the constants for hydrogen sulfide have the following values:

$$K_p = 3.6 \times 10^{-2} \ M \ atm^{-1} \quad K_1 = 2.9 \times 10^{-7} \ M \quad K_2 = 1.1 \times 10^{-10} \ M$$

and the solubility product of cadmium sulfide is $4.8 \times 10^{-23} \ M^2$. Let us calculate the maximum hydrogen ion concentration such that 99.8% of 0.010 M Cd^{++} will have precipitated as the sulfide.

The concentration of cadmium ion remaining in solution at equilibrium is required to be

$$[Cd^{++}] = \frac{0.2}{100} \times 0.010 = 2.0 \times 10^{-5} \ M$$

This must satisfy the solubility product, so that sulfide ion concentration is

$$[S^{--}] = \frac{K_{sp}}{[Cd^{++}]} = \frac{4.8 \times 10^{-23}}{2.0 \times 10^{-5}} = 2.4 \times 10^{-18} \ M$$

The concentration of sulfide must also satisfy the ionization equilibrium of hydrogen sulfide. If the solution is saturated with hydrogen sulfide gas at 1 atm pressure, equation (2.2) is applicable:

$$K_{ip} = [H^+]^2 [S^{--}] = K_p K_1 K_2 = 1.12 \times 10^{-18} \ M^3$$

[2] If, in the hydrolysis of thioacetamide, hydrogen sulfide is being consumed by precipitation as fast as it is produced, the balance between production and consumption will keep [H₂S] very small but roughly constant during precipitation. The ion product relation [equation (2.2)] will still hold, but its numerical value will be smaller than that for a saturated solution of the gas.

Into this expression we introduce the prescribed sulfide ion concentration and solve for the hydrogen ion concentration:

$$[H^+]^2 = \frac{K_{ip}}{[S^{--}]} = \frac{1.12 \times 10^{-18}}{2.4 \times 10^{-18}} = 0.47$$

whence $[H^+] = 0.69\ M$.

This calculated value of the hydrogen ion concentration is not far from the experimental range of 0.1–0.3 M. Exact agreement is scarcely to be expected, if only because of uncertainties in the various constants. The solubility product, in particular, is for stable, crystalline CdS. Freshly precipitated sulfides are not yet at true equilibrium and are commonly more soluble, so that a larger effective solubility product ought to be used. This would reduce the calculated value of the hydrogen ion concentration.

Suppose the solution also contains Fe^{++} and Zn^{++} ions at 0.010 M concentration. Let us see if calculations indicate whether or not their sulfides will precipitate under the given conditions. Iron(II) sulfide has a much larger solubility product than cadmium sulfide, namely, $6.0 \times 10^{-15}\ M^2$ at 100°. The ion concentration product

$$[Fe^{++}][S^{--}] = 0.010 \times 2.4 \times 10^{-18} = 2.4 \times 10^{-20}$$

is therefore much less than K_{sp}, the equilibrium value. The solution is unsaturated in FeS, and it cannot precipitate. Zinc sulfide has a solubility product that is much closer to that of cadmium sulfide; a reasonable guess for the solubility product of freshly precipitated ZnS is a value larger than 1×10^{-19}, so that here too the ion concentration product of 2.4×10^{-20} is below the equilibrium value, and no precipitation will occur.

2.7 Precipitation of Hydroxides. Control of the Hydroxide Ion Concentration

Insoluble hydroxides, like sulfides, fall into two broad classes, those in one being more soluble than those in the other. The hydroxides of manganese(II) and chromium(III) are representative examples[3]: their solubility products are, respectively, 1.6×10^{-13} and 7×10^{-31}. By controlling the hydroxide concentration at a low level, we can establish the conditions

$$K_{sp} > [Mn^{++}][OH^-]^2 \quad \text{and} \quad K_{sp} < [Cr^{3+}][OH^-]^3$$

which determine that manganese hydroxide will not precipitate, whereas chromium hydroxide will. If the hydroxide ion concentration is allowed to become too high, both hydroxides will come down; if too low, neither will precipitate.

---

[3] The fact that some insoluble hydroxides like $Cr(OH)_3$ are more properly regarded as hydrated oxides $(Cr_2O_3 \cdot xH_2O)$ does not affect the following arguments because either form is in equilibrium with the same ions in solution.

Control of the hydroxide ion concentration is established by means of a buffer mixture of a weak base and its salt, most commonly ammonia and ammonium chloride. The ionization equilibrium of the base

$$NH_3 + H_2O \rightleftharpoons NH_4^+ + OH^-$$

is characterized by the basicity constant

$$K_b = \frac{[NH_4^+][OH^-]}{[NH_3]} \tag{2.4}$$

The numerical value of this is such that in 0.10 M NH_3 the hydroxide ion concentration is about 0.0013 M, sufficiently large to precipitate hydroxides of both classes.

The salt in the buffer mixture supplies the common ion NH_4^+, which depresses ionization of the base, lowering the hydroxide ion concentration. The value of $[OH^-]$ will evidently depend on the ratio of salt to base in the buffer, and it can be adjusted so that separation of the hydroxides is achieved.

Not only does the buffer set a certain concentration of hydroxide ion but it also tends to maintain this level as small amounts of hydroxide are withdrawn to form a precipitate. This action requires that the concentrations of both base and salt be high, so that small shifts in the equilibrium caused by removal of hydroxide do not affect the ratio appreciably. Let the large concentrations of base and salt be represented by large rectangles:

$$[OH^-] = K_b \frac{\boxed{NH_3}}{\boxed{NH_4^+}}$$

As OH^- ions are removed, more ammonia ionizes to restore hydroxide, resulting in the formation of more ammonium ion but little change in $[OH^-]$:

$$[OH^-] = K_b \frac{\boxed{NH_3}}{\boxed{NH_4^+}}$$

2.8 Numerical Calculations
for the Precipitation of Hydroxides

To test the principles discussed in the preceding section, let us see if we can verify by numerical calculation that a buffer that is 1.0 M in NH_3 and 3.0 M in NH_4Cl can effect a separation of aluminum from magnesium. In view of the high concentration of base and salt and the low degree of ionization of the base under these conditions, we can use $[NH_3] = 1.0 \ M$ and $[NH_4^+] = 3.0 \ M$ in equation (2.4).[4] On that basis and with the known value of $K_b = 1.8 \times 10^{-5} \ M$, we can calculate the hydroxide ion concentration

$$[OH^-] = 1.8 \times 10^{-5} \times \frac{1.0}{3.0} = 6.0 \times 10^{-6} \ M$$

If both metal ions are initially at 0.010 M concentration, we then have the conditions

$$Mg(OH)_2: \quad K_{sp} = 6.3 \times 10^{-10} > 0.010 \times (6.0 \times 10^{-6})^2$$

$$Al(OH)_3: \quad K_{sp} = 3.1 \times 10^{-34} < 0.010 \times (6.0 \times 10^{-6})^3$$

We therefore expect aluminum hydroxide, but not magnesium hydroxide, to precipitate. From another point of view, the concentrations of metal ions that could remain in solution in equilibrium with $[OH^-] = 6.0 \times 10^{-6} \ M$ are

$$[Al^{3+}] = \frac{3.1 \times 10^{-34}}{(6.0 \times 10^{-6})^3} = 8.6 \times 10^{-19} \ M$$

an exceedingly small value, and

$$[Mg^{++}] = \frac{6.3 \times 10^{-10}}{(6.0 \times 10^{-6})^2} = 17.5 \ M$$

an impossibly large value.

Our prediction that the ammonia–ammonium chloride buffer will separate magnesium from aluminum is, for the most part, substantiated by experiment. Insoluble hydroxides are gelatinous and adsorptive precipitates, so that they tend to drag foreign ions from solution. Some magnesium will therefore usually coprecipitate with aluminum hydroxide. Moreover, freshly precipitated hydroxides, like sulfides, are not in strict equilibrium with the solution; they become less soluble as they age. Our calculations used solubility products of the active or freshly precipitated compounds, and these are only approximate values. Not infrequently, the first compound precipitated when a strong base is added to

---

[4] Accurate expressions for these concentrations are

$$[NH_3] = 1.0 - [OH^-] + [H^+]$$

and

$$[NH_4^+] = 3.0 + [OH^-] - [H^+]$$

a salt is not a pure hydroxide but a basic salt like $Ni(OH)Cl$ or $Co(OH)Cl \cdot 4Co(OH)_2 \cdot 4H_2O$. As more strong base is added, these precipitates gradually change to the crystalline hydroxides or hydrated oxides. Neither nickel(II) nor cobalt(II) hydroxides would be precipitated by the ammonia–ammonium chloride buffer because of ammonia complex formation (discussed in Sec. 2.10).

2.9   Separations by Control of Carbonate and Chromate Ion Concentrations

Buffer action can also be used to control other ion concentrations. Magnesium carbonate is more soluble than most carbonates and can be prevented from precipitating by controlling the carbonate ion concentration. For this purpose the equilibrium

$$NH_4^+ + CO_3^{--} \leftrightharpoons NH_3 + HCO_3^-$$

present in a solution of ammonium carbonate is suitable. In 0.10 M ammonium carbonate alone, the carbonate ion concentration is 0.0067 M, high enough to precipitate magnesium carbonate if the magnesium ion concentration is much above 0.010 M. A mixture of high concentrations of ammonium chloride, ammonia, and ammonium carbonate establishes a suitably low concentration of carbonate ion to prevent the precipitation of magnesium carbonate while allowing precipitation of more insoluble carbonates like $CaCO_3$. The ammonia–ammonium chloride buffer ensures that as carbonate is removed to form $CaCO_3$, more carbonate will be generated to maintain approximately the same concentration. Moreover, the buffer prevents precipitation of magnesium hydroxide, which might occur if the solution were too basic. More details are given in Sec. 10.3. Numerical calculations for this system are unreliable because ammonium carbonate solutions always contain considerable quantities of ammonium carbamate, $NH_4CO_2NH_2$.

Separation of chromates is ensured by use of an acid buffer. Barium chromate, the least soluble, can then be precipitated without contamination by strontium or calcium chromates when the chromate ion concentration is regulated at a low level:

$$K_{sp} < [Ba^{++}][CrO_4^{--}] \quad \text{but} \quad K_{sp} > [Sr^{++}][CrO_4^{--}]$$

Chromate is controlled by its equilibrium with dichromate

$$2CrO_4^{--} + 2H^+ \leftrightharpoons Cr_2O_7^{--} + H_2O$$

with equilibrium constant

$$K = \frac{[Cr_2O_7^{--}]}{[CrO_4^{--}]^2[H^+]^2} = 3.3 \times 10^{14}$$

The ratio $[Cr_2O_7^{--}]/[CrO_4^{--}]^2$ evidently depends on the hydrogen ion concentration, being 3.3 at pH 7, 3.3×10^8 at pH 3, and 3.3×10^{-4} at pH 9. The pH can be set by means of a weak acid–salt type of buffer, such as ammonium acetate–acetic acid. By regulating the hydrogen ion concentration, we fix the chromate ion concentration at such a level that only barium chromate precipitates. Further details and numerical calculations are given in Secs. 9.13 and 10.4.

2.10 Control of Metal Ion Concentrations by Complex Formation

Up to this point we have discussed only separations based on control of the concentration of an anion. We can tie up metal ions in weakly dissociated complexes and bring about separations in this way. It is convenient, for example, to separate sulfide, chloride, bromide, and iodide ions from other anions by precipitating them as silver salts. Some control of the silver ion concentration is desired because other anions—for example, PO_4^{3-} and SO_3^{--}—may form precipitates if the silver concentration is high. The silver ion concentration is controlled by the equilibrium

$$Ag^+ + 2NH_3 \leftrightharpoons Ag(NH_3)_2^+ \qquad \beta_2 = 1.6 \times 10^7 \ M^{-2}$$

To keep it low, an excess of ammonia must be present; but if too large an excess is used, silver chloride will not precipitate completely. The reagent for precipitation therefore includes not only silver nitrate and ammonia but also ammonium carbonate, which contributes the additional equilibrium

$$NH_4^+ + CO_3^{--} \leftrightharpoons NH_3 + HCO_3^-$$

If by chance too much ammonia is present, some will be removed by combination with hydrogen carbonate ion. On the other hand, if more ammonia is required, it can be generated by combination of ammonium and carbonate ions. With a regulated concentration of ammonia, the concentration of silver ion is controlled so that only the least soluble silver salts precipitate.

2.11 Numerical Calculations for Complex Formation

It is observed that aluminum ion can be separated from zinc ion by a combination of ammonia and ammonium chloride. The hydroxide ion concentration of a solution that is $0.50 \ M$ in both reagents is numerically equal to K_b of ammonia:

$$[OH^-] = K_b \frac{[NH_3]}{[NH_4^+]} = K_b \times \frac{0.5}{0.5} = 1.8 \times 10^{-5} \ M$$

This is sufficiently high to precipitate both $Zn(OH)_2$ and $Al(OH)_3$ from $0.010 \ M$ solutions of their cations if no other equilibria are involved:

$Zn(OH)_2$: $[Zn^{++}][OH^-]^2 > K_{sp}$ or $0.01 \times (1.8 \times 10^{-5})^2 > 1.1 \times 10^{-16}$

$Al(OH)_3$: $[Al^{3+}][OH^-]^3 > K_{sp}$ or $0.01 \times (1.8 \times 10^{-5})^3 > 3.1 \times 10^{-34}$

In fact, zinc forms a weakly dissociated ammonia complex $Zn(NH_3)_4^{++}$ with overall formation constant

$$\beta_4 = \frac{[Zn(NH_3)_4^{++}]}{[Zn^{++}][NH_3]^4} = 9 \times 10^9$$

To simplify the calculations, we suppose that virtually all the zinc goes into the complex

$$[Zn(NH_3)_4^{++}] = 0.010 - [Zn^{++}] \cong 0.010 \ M$$

The equilibrium concentration of ammonia is the total minus that used to form the complex

$$[NH_3] = 0.50 - (4 \times 0.010) = 0.46 \ M$$

Then

$$[Zn^{++}] = \frac{0.010}{9 \times 10^9 \times (0.46)^4} = 2.5 \times 10^{-11} \ M$$

This is too low to satisfy the solubility product expression

$$2.5 \times 10^{-11} \times (1.8 \times 10^{-5})^2 < 1.1 \times 10^{-16} = K_{sp}$$

and zinc hydroxide will not precipitate.

Zinc ion can also form a hydroxo complex, the zincate ion $Zn(OH)_4^{--}$, with overall formation constant

$$\beta_4 = \frac{[Zn(OH)_4^{--}]}{[Zn^{++}][OH^-]^4} = 3 \times 10^{15}$$

We might suppose at first glance that, because the hydroxo complex has the larger formation constant, most of the zinc would be in this form. But the buffer maintains a relatively low (1.8×10^{-5}) concentration of hydroxide ion and a high concentration (near $0.5 \ M$) of ammonia molecules. To make a comparison taking these concentrations into account, we could eliminate the zinc ion concentration by taking the ratio of the two formation constants. After some rearrangement, we obtain for the ratio of concentrations of the two complexes the expression

$$\frac{[Zn(NH_3)_4^{++}]}{[Zn(OH)_4^{--}]} = \frac{\beta_4 \ of \ Zn(NH_3)_4^{++} \times [NH_3]^4}{\beta_4 \ of \ Zn(OH)_4^{--} \times [OH^-]^4}$$

$$= \frac{9 \times 10^9 \times (0.5)^4}{3 \times 10^5 \times (1.8 \times 10^{-5})^4} = 2 \times 10^{12}$$

This large value shows that practically all the zinc is in the form of the ammonia complex. The hydroxo complex is usually formed by adding an excess of a strong base like sodium hydroxide to a zinc salt.

2.12   Separation by Selective Redox Reactions

Two ions may both be capable of being oxidized or reduced, yet it is frequently possible to find a reagent that will affect only one of them. Less commonly, a reagent capable of attacking both may be prevented from acting on one by proper choice of concentration or acidity.

Iodide and bromide ions, for example, can both be oxidized to the free halogens, but only iodide is oxidized by dilute nitrous acid:

$$2e^- + I_2 \rightarrow 2I^- \qquad\qquad E° = 0.535 \text{ V}$$

$$e^- + H^+ + HNO_2 \rightarrow NO + H_2O \qquad E° = 0.96 \text{ V}$$

$$2e^- + Br_2 \rightarrow 2Br^- \qquad\qquad E° = 1.065 \text{ V}$$

After removal of the iodine, bromide can be oxidized by permanganate, a stronger oxidizing agent than nitrous acid:

$$5e^- + 8H^+ + MnO_4^- \rightarrow Mn^{++} + 4H_2O \qquad E° = 1.51 \text{ V}$$

Although permanganate is also capable of oxidizing chloride,

$$2e^- + Cl_2 \rightarrow 2Cl^- \qquad E° = 1.36 \text{ V}$$

the reaction is slow enough so that bromide can be detected and eliminated without much loss of chloride.

Selective reduction is used to separate copper from cadmium. The reducing agent dithionite ion, $S_2O_4^{--}$, contains sulfur in an oxidation state of $3+$ and is oxidized to sulfite ion, SO_3^{--}, in ammoniacal solution. Dithionite will reduce copper(II) to metallic copper but does not affect cadmium:

$$Cu(NH_3)_4^{++} + 2e^- \rightarrow Cu + 4NH_3 \qquad\qquad E° = -0.06 \text{ V}$$

$$SO_3^{--} + 4NH_4^+ + 2e^- \rightarrow S_2O_4^{--} + 4NH_3 + 2H_2O \qquad E° = -0.56 \text{ V}$$

$$Cd(NH_3)_4^{++} + 2e^- \rightarrow Cd + 4NH_3 \qquad\qquad E° = -0.61 \text{ V}$$

Once copper is out of the way, it is possible to identify cadmium in the solution by precipitating yellow CdS. If copper were present, black CuS would form, obscuring the yellow sulfide.

2.13   Separation by Physical Methods. Distribution Equilibria

Several methods make use of physical means to bring about a separation. For example, differences in the degree of absorption of ions by paper and in the solubility of their compounds in a solvent are capitalized on in paper chromatography. While some of these methods are very discriminating and powerful, we shall deal here only with the simple process of distribution of a solute between immiscible liquids.

Carbon disulfide and water do not dissolve in each other to an appreciable extent; they are immiscible. Iodine is more soluble in carbon disulfide than in water. When some water containing dissolved iodine is shaken with carbon disulfide, the iodine distributes itself between the two liquids, with the bulk of it going into the carbon disulfide layer. The equilibrium condition is expressed by the ratio

$$K_D = \frac{[I_2]_{CS_2}}{[I_2]_{H_2O}} \tag{2.5}$$

Distribution between immiscible solvents is used in two ways in qualitative analysis. Identification reactions can be based on the extraction of a colored form of a substance from water solution into another solvent. The violet color of iodine and the orange color of bromine in carbon disulfide or carbon tetrachloride are examples. Extraction can also be used to remove large quantities of a troublesome ion. Iron in a ferrous alloy is present in such overwhelming quantity that it makes tests for other metals difficult. If the iron is converted to iron(III) chloride, it can be extracted by ether, leaving the other metal ions behind in the water layer.

As an example of a numerical computation based on the distribution ratio, let us calculate the amount of bromine extracted when 2.0 ml of water containing 4.0 mg of bromine is shaken with 1.0 ml of carbon tetrachloride. The distribution ratio is

$$K_D = \frac{[Br_2]_{CCl_4}}{[Br_2]_{H_2O}} = 22.7 \text{ at } 25°C$$

Let x be the milligrams of bromine extracted. Then $4.0 - x$ must be the milligrams of bromine left in the water layer. These expressions are converted to concentrations by dividing by the volumes. Then

$$\frac{x \text{ mg}/1.0 \text{ ml}}{(4.0 - x) \text{ mg}/2.0 \text{ ml}} = 22.7 = \frac{2x}{4 - x}$$

which gives $x = 90.8/24.7 = 3.7$ mg extracted, leaving 0.3 mg in the water layer. Although the distribution ratio was expressed as the ratio of molar concentrations, we have used the ratio of concentrations in milligrams per milliliter. The conversion from one unit to the other is the same in numerator and denominator.

EXERCISES

Values of solubility products, acidity and basicity constants, and ion products are given in Appendix A.

2.1.   What is meant by "precipitation from homogeneous medium?" What are the advantages of this technique over rapid precipitation? Give two illustrations.

2.2. Urea, $CO(NH_2)_2$, slowly hydrolyzes in boiling water to ammonia and carbon dioxide. What use might be made of this reaction in separating ions?

2.3. Consider the following relationships:

$$[Fe^{++}][S^{--}] = K_{sp}; \quad [Fe^{++}][S^{--}] > K_{sp}; \quad [Fe^{++}][S^{--}] < K_{sp}$$

(a) Which one is the condition for an unsaturated solution?
(b) Which one indicates a stable state of the system?
(c) Which one must be established before precipitation of FeS will occur?

2.4. Represent by chemical equations and equilibrium constant expressions the equilibria in a saturated solution of hydrogen sulfide. Explain on the basis of these expressions how insoluble metal sulfides can be separated into two groups.

2.5. Discuss the effect of adding ammonium nitrate to the equilibrium existing in aqueous ammonia. Why is the resulting mixture called a "buffer?" How can such a mixture be used in separating ions?

2.6. What combination of reagents is used to control the carbonate ion concentration? Write chemical equations and equilibrium constant expressions. What would be the disastrous consequences of too high a carbonate ion concentration? Of one that is too low? How is buffer action used in the control?

2.7. Write a chemical equation for the chromate–dichromate equilibrium, and explain how the chromate ion concentration is controlled. The precipitation of barium chromate releases hydrogen ion, as shown by

$$2Ba^{++} + Cr_2O_7^{--} + H_2O \rightarrow 2BaCrO_4(s) + 2H^+$$

How is this compensated for?

2.8. Iron(III) ion forms neither an ammonia complex nor a hydroxo complex. Aluminium(III) ion forms only the hydroxo complex, and zinc(II) ion forms both. Outline a scheme whereby the three ions could be separated from each other using only aqueous ammonia and sodium hydroxide as reagents.

2.9. Sulfite ion interferes with the identification of carbonate ion, for both ions react with acid to give volatile acidic oxides. Before making a test for carbonate, sufite must be removed by treatment with hydrogen peroxide. The possible half-reactions are

$$3H_2O + SO_3^{--} + 4e^- \rightarrow S + 6OH^- \qquad E° = -0.66 \text{ V}$$
$$H_2O + SO_4^{--} + 2e^- \rightarrow SO_3^{--} + 2OH^- \qquad E° = -0.93 \text{ V}$$
$$HO_2^- + H_2O + 2e^- \rightarrow 3OH^- \qquad E° = +0.88 \text{ V}$$
$$O_2 + H_2O + 2e^- \rightarrow HO_2^- + OH^- \qquad E° = -0.08 \text{ V}$$

Will sulfite be oxidized or reduced by hydrogen peroxide (which occurs as HO_2^- in basic solution)? Derive a complete equation for the reaction and calculate the total $E°$ value. Do you expect it to be positive or negative?

2.10. Solid silver nitrate is added gradually to a solution that contains 0.0010 mole of chromate and 0.020 mole of chloride per liter. Assume that the volume remains constant during the addition.

(a) Which will precipitate first, AgCl or Ag_2CrO_4? Calculate the concentration of silver ion when each precipitate first begins to form.

(b) What will be the concentration of chloride ion when silver chromate first begins to precipitate?

(c) What percentage of the chloride has precipitated when silver chromate begins to form?

2.11. It is proposed to separate manganese(II) and copper(II) by fractional precipitation of their sulfides in a solution of controlled acidity. Sulfide is generated by the hydrolysis of thioacetamide. The metal ions are both initially at 0.10 M concentration.

(a) Calculate the concentration of sulfide ion required to form each precipitate. Which will precipitate first?

(b) What will be the concentration of copper ion when manganese sulfide starts to precipitate? What percentage of the copper has precipitated at this point?

2.12. The pH at which zinc sulfide starts to precipitate from 0.010 M zinc nitrate solution saturated with hydrogen sulfide at 25°C and 1 atm pressure is 1.15. To what value of the solubility product of zinc sulfide does this correspond? Why is it different from the one given in Appendix A.2?

2.13. Find the approximate concentration of zinc ion left unprecipitated if a 0.10 M solution of a zinc is made 1.0 M in monochloracetic acid and 0.50 M in the sodium salt of the acid and is saturated with hydrogen sulfide at 1 atm pressure and 25°C. The solubility product of freshly precipitated zinc sulfide is about $1 \times 10^{-21} \ M^2$.

2.14. Calculate the number of grams of solid ammonium chloride that must be added to 100 ml of 0.050 M ammonia if iron(II) hydroxide is not to precipitate when the mixture is added to 100 ml of 0.020 M iron(II) chloride solution. Assume a final volume of 200 ml, and do not neglect the changes in concentration that result from mixing the two solutions; for example, $[Fe^{++}] = 0.010 \ M$, not 0.020 M.

2.15. Can a buffer that is 0.10 M in triethanolamine and 0.50 M in the nitrate salt of this base separate Mn^{++} from Cr^{3+} by precipitation of one hydroxide and not the other? Prove by appropriate calculations. Assume 0.050 M solutions of the metal ions and suppose that for a successful separation not more than 0.010 mg Cr per ml must be left in solution.

2.16. If a small amount of mercury(II) nitrate is added to a solution that is 0.50 M in chloride ion and 0.0010 M in iodide ion, what percentage of the mercury will be in the form of the chloride complex? Overall formation constants are $1.2 \times 10^{15} \ M^{-4}$ for $HgCl_4^{--}$ and $1.9 \times 10^{30} \ M^{-4}$ for HgI_4^{--}.

2.17. How many moles of potassium iodide must be added to a liter of mixture that is 1.0 M in hydrogen ion and 0.10 M in mercury(II) nitrate if mercury(II) sulfide is to be kept from precipitating when the mixture is saturated with hydrogen sulfide at 25°C and 1 atm? The overall formation constant of HgI_4^{--} is $1.9 \times 10^{30} \ M^{-4}$.

2.18. At 25°C the equilibrium constant for the reaction

$$Al(OH)_3(s) + OH^- \leftrightharpoons Al(OH)_4^-$$

is 0.295. Calculate the solubility of aluminum hydroxide in sodium hydroxide solution which is initially 0.50 M. Combine this equilibrium constant with the value of the solubility product of aluminum hydroxide given in

Appendix A.2 to get the overall formation constant β_4 for the hydroxo complex.

2.19. A solution of iodine containing 0.335 g of the element in 25 ml of carbon tetrachloride is shaken with 250 ml of water. The aqueous layer is found to contain 0.035 g of iodine. Calculate the distribution ratio $K_D = [I_2]_{CCl_4}/[I_2]_{H_2O}$.

2.20. The distribution ratio $[Br_2]_{CCl_4}/[Br_2]_{H_2O}$ is 28 if the water layer is acidified to suppress hydrolysis of bromine.

   (a) If 5.0 ml of water containing 2.0 mg of bromine is shaken with 2.0 ml of carbon tetrachloride, how much bromine is extracted?

   (b) How much bromine would be removed if another 5.0-ml portion of this aqueous solution was extracted with two successive 1.0-ml portions of carbon tetrachloride? What conclusion can be drawn regarding the efficiency of a single extraction as against multiple extraction with the same total volume of liquid?

Techniques

THREE

The Laboratory Work

in Qualitative Analysis

3.1 Preliminary Work

The following work should require at most one period.

(*a*) *Check in.* After you have been assigned a desk, check the equipment in it against the apparatus list, and replace missing items. Inspect all glass and porcelain ware for chips, cracks, or bad scratches. If the apparatus and desk are dirty, some housecleaning is in order.

(*b*) *Examine the reagent kit.* Commonly used reagents may be supplied in a kit on your desk. Glass-stoppered bottles are usually used for the concentrated acids, polyethylene bottles for the strong bases, and amber bottles for reagents such as $AgNO_3$, $KMnO_4$, and H_2O_2 that are affected by light. Arrange the bottles in some systematic order—for example, alphabetically—and keep them that way. A satisfactory arrangement for the larger bottles is shown below. The most

Back

NH_3 6 *M*	KOH 0.5 *M*	H_2SO_4 18 *M*	HOAc 6 *M*	HNO_3 6 *M*	HNO_3 16 *M*
NH_3 15 *M*	NaOH 6 *M*	Thio-acet-amide	HCl 2 *M*	HCl 6 *M*	HCl 12 *M*

Front

commonly used reagents are in front. The ammonia solutions are separated as far as possible from nitric and hydrochloric acids, because ammonia and acid fumes combine to give unsightly white deposits of ammonium salts on the bottles.

If your reagent kit is not already labeled and filled, your instructor will supply instructions for this purpose.

(*c*) *Make six glass rods.* Cut a rod with a diameter of 2 mm into suitable lengths as shown in Fig. 3.1. Fire polish each end most carefully, because sharp edges will scratch test tubes and make them susceptible to breakage at inopportune times. Handles may be made on some of the rods as illustrated.

(*d*) *Calibrate four or more capillary pipets.* You have been supplied with disposable Pasteur pipets with capillary tips 50 mm long. These are used for withdrawing and transferring solutions from test tubes and centrifuge tubes. The small end takes up less room in a narrow tube than the ordinary medicine dropper would. Calibrate each pipet by counting the number of drops it delivers into a 10-ml graduated cylinder. Deliver at least 2 ml of water. Label the shank of the pipet with the number of drops per milliliter; it should deliver between 20 and 40.

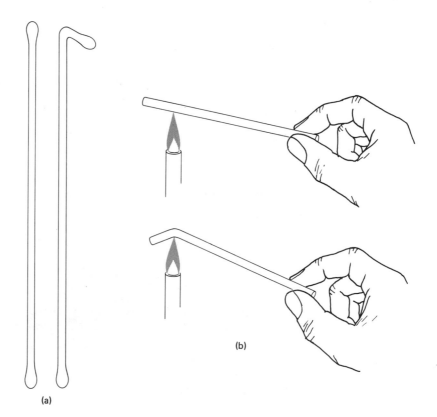

(b)

(a)

Fig. 3.1. Glass rods. (a) Actual size. (b) Making a handle on a rod.

3.2   General Remarks About the Laboratory Work

The immediate goal of your laboratory work is to determine what ions are in a particular sample. You may be answering practical questions: Is this solid common salt or rat poison? Does this detergent contain phosphate? The inquisitive student, while pursuing this mundane goal, also seeks to understand the procedures and to learn some inorganic chemistry. Moreover, the nature of the experimental work provides an opportunity to learn good laboratory habits.

You will find the laboratory periods more worthwhile and more enjoyable if you plan the work ahead of time, organize your desk space for efficient operation, and develop habits that ensure cleanliness and guard against contamination. You will benefit most if you give careful thought to these matters and work out your own procedures. The following remarks are meant only to be suggestions, not prescriptions. The exact arrangement of your desk, for example, will depend on the area available, the position of sink and gas outlets, and shelf space. One such arrangement is shown in Fig. 3.2. The reagents are easily visible and accessible, yet they are not in the way. Flames and the hot water bath are kept away from the reagent racks because several reagents are adversely affected by heat. After using a pipet, rinse it twice with distilled water from beaker *A* and once from beaker *B*. Pipets and rods can be stored in *B* to keep them off the desk top. If pipets are stored tip down as shown, solution cannot flow back into the rubber bulb.

Safety precautions also deserve consideration. It is common sense to protect your eyes with goggles or glasses and your clothes with a coat or apron. Many

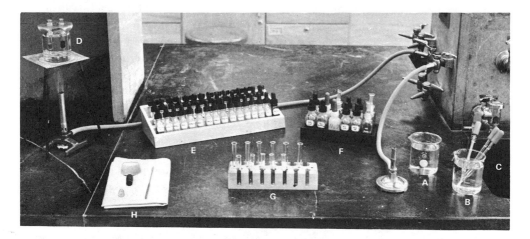

Fig. 3.2. The desk top. A possible arrangement of the working space. A, B, beakers containing distilled water, rods, and pipets; C, sink; D, water bath; E, large reagent bottles in rack; F, small reagent bottles in rack; G, test tube rack; H, folded towel on which rest crucible, casserole, spatula, etc.

chemicals are poisonous, so it makes sense not to mix food with qualitative analysis. Likewise, since many chemicals are corrosive, it is a good rule to wash off with plenty of water any chemical that is spilled on your skin.

3.3   The Techniques of Semimicro Qualitative Analysis

While some of these techniques are peculiar to work using small-scale equipment, most are of general application and, with some adaptations, are used in other chemistry courses. Read through the description of these techniques before starting the laboratory work. Then reread them as they are referred to in the directions for analysis.

T-1.   Sampling

It is essential that the sample taken for analysis from a large quantity of material be a representative one; that is, it must contain all the constituents of the material. For a solution, it is necessary only that it be mixed well before the sample is removed. If it contains suspended matter, shake it thoroughly and withdraw the sample quickly before the solids have time to settle. Solid materials may be far from homogeneous. Even after they have been carefully mixed, the constituents may separate after the mixture is bottled if they differ considerably in density. Examine the material under a hand lens for evidence of nonhomogeneity and mix it thoroughly before taking a sample.

T-2.   Dissolving the Sample

Only the simpler situations are considered here; more details are given in later chapters. The most desirable solvent is water. Test the solubility of a small amount (less than 20 mg) in several drops of water. If solution does not take place rapidly at room temperature after adequate stirring, add some more water and stir again. If this has no effect, try warming it for several minutes in the water bath. If there is a residue insoluble in water, test its solubility in 6 M nitric or hydrochloric acid. Allow plenty of time for solution to take place. Some salts hydrolyze extensively in water to give precipitates; it is often easier to dissolve them directly in acid than it is to try to dissolve the precipitate that forms by hydrolysis. Concentrated acids and aqua regia (3:1 HCl–HNO_3) can sometimes be used to advantage but should be avoided if possible, for the excess acid will have to be neutralized or removed by evaporation. Certain salts are also remarkably insoluble in concentrated acids.

T-3.   Measuring Quantity

(a) The quantity of *solutions* is measured in drops or milliliters. One standard drop is 0.05 ml; thus there are 20 such drops per milliliter. It is sufficient to assume for most work that the droppers in the reagent kit deliver drops of this size. When more accurate measurement of volume is required, use calibrated capillary pipets.

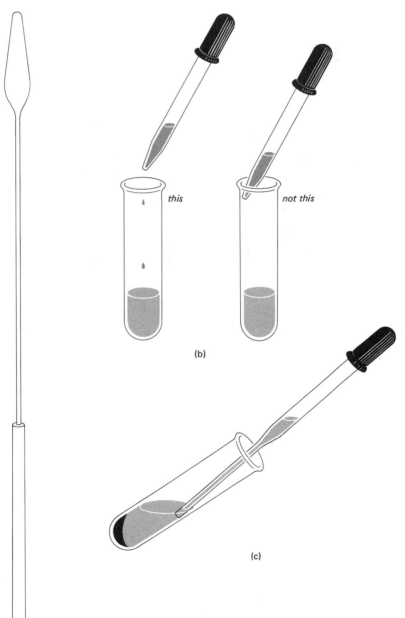

this   *not this*

(b)

(c)

(a)

Fig. 3.3. Techniques of measurement and pipetting. (a) Semimicro spatula (actual size). (b) Addition of reagents. Correct and incorrect techniques. (c) Separation of solution from precipitate with a capillary pipet.

(*b*) The quantity of *solids* is measured with a balance. A triple beam balance sensitive to 10 mg (0.01 g) is satisfactory for most work. Weigh by difference: find the weight of an empty test tube or micro beaker and weigh the sample into it. When measurement of an exact quantity is not important, a spatula may be used. A heaping spatula-full, for a spatula of the size pictured in Fig. 3.3, contains about 0.10–0.15 g of inorganic salts.

T-4. Adding Reagents

It is imperative that reagents be preserved against contamination. Droppers of reagent bottles should always be held above tubes and other vessels and not be allowed to touch them (Fig. 3.3). Sometimes, particularly in neutralization, it is desirable to add less than one drop of reagent. Give the dropper bulb the slightest squeeze, and remove the fraction of a drop that emerges first to a clean stirring rod and then to the solution.

T-5. Mixing

Because of the narrow bore of test tubes and centrifuge tubes, this is one of the most critical and difficult operations of semimicro analysis. Yet the first rule of analysis must be to mix the solution thoroughly before drawing any conclusions. If you close the tube with a finger or cork while you shake it to mix the sample, you will very likely contaminate the solution. The following techniques are better.

(*a*) If the solution fills no more than half the test tube, you can mix it well by grasping the top firmly between the thumb and forefinger of one hand and flicking the bottom of the tube with the forefinger of the other hand. This will not work as well with centrifuge tubes because of their tapered bottoms.

(*b*) If the test tube is more than half full, it is best to mix by pouring the contents from one tube to another or from tube to casserole or micro beaker.

(*c*) You can also mix a solution efficiently by sucking up a portion of it with a capillary pipet (not all the way up to the bulb) and expelling it at the bottom of the tube. Repeat at least twice.

(*d*) You may stir the solution with a glass rod, using a combination of up and down and circulatory motions, but this is the least efficient of the four techniques. Glass rods are more useful for such purposes as testing with litmus or loosening a cake of precipitate than for stirring.

T-6. Heating Solutions

Solutions in casseroles can be heated directly over a free flame. This is unsatisfactory for solutions in test tubes or centrifuge tubes because of the tendency of steam bubbles to form at the bottom of the tube and expel solution as they rise and expand. Always heat solutions in tubes by setting them in a hot water bath. A simple bath can be made from a 250-ml beaker and a metal cover containing three or more holes [Fig. 3.4(a)]. If a rubber band is wrapped around the top of the tube, it will be easier to lift out.

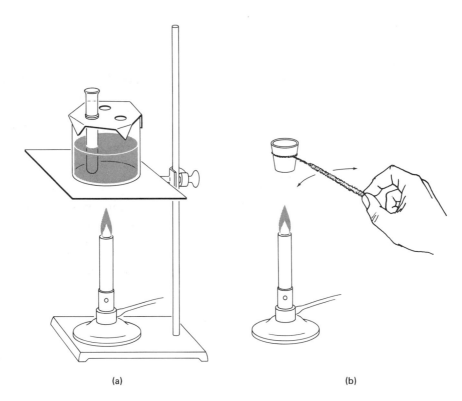

(a)   (b)

Fig. 3.4. Techniques for heating and evaporation. (a) Water bath. (b) Evaporation of a few drops in a crucible. Note holder made from a pipe cleaner.

T-7.   Evaporating Solutions

Casseroles are the most convenient vessels for evaporating several milliliters of solution. An inch or two of heavy rubber tubing slipped halfway up the handle will make the casserole easier to handle when it is hot. Micro crucibles can be used for the evaporation of a few drops. Construct a simple holder [Fig. 3.4(b)] by burning off half the fuzz from a pipecleaner and twisting the bare wire into a loop. The crucible should fit two-thirds of the way down into the loop.

Evaporations must be carried out so that material will not be lost by spattering. Usually it is undesirable to overheat a residue left after complete evaporation, for the residue is often volatile or it changes to a less soluble form on baking. The following techniques will help you avoid these difficulties.

(*a*) Use a microburner and adjust it to give a flame not more than $\frac{1}{2}$ inch high. If necessary, use a screw clamp on the burner tubing to make this adjustment. This will be referred to in the analytical procedures as a *microflame*.

(*b*) During evaporations in casseroles, keep the contents agitated by a swirling motion. As dryness approaches, the center of the bottom will go dry first. Keep it moist as long as possible by swirling and tilting motions. *Before complete dryness is reached*, withdraw the casserole from the flame, and let the heat of the dish complete the evaporation.

T-8. Centrifugation

This is much quicker than filtration for the separation of small quantities. A centrifuge subjects an object to a force far in excess of that of gravity. If d is the distance from the axis of rotation to the precipitate, the force acting on the precipitate is proportional to d multiplied by the square of the speed of rotation. When d is 11 cm and the speed of rotation is 1650 rpm, the force is about 330 times that of gravity. The precipitate will settle 330 times faster in the centrifuge than it would if allowed to settle in a test tube placed in a rack. The following points should be observed in using centrifuges.

(a) *The revolving head must be carefully balanced.* Always balance the tube containing the sample with another one containing an equal volume of water; place them opposite each other in the head. Balance centrifuge tube against centrifuge tube, and test tube against test tube. An unbalanced centrifuge will wobble or "walk" on the desk. This may damage the centrifuge and endanger bystanders.

(b) *Centrifuge only a few seconds at top speed.* You share the centrifuge with others; give them a chance to use it.

(c) *Test tubes are satisfactory for most centrifugations.* They are preferable to centrifuge tubes in general use because solutions are more easily mixed in tubes with a wide bottom. If the precipitate is small or does not pack down tightly in a test tube, a cleaner separation is obtained in a centrifuge tube.

T-9. Separating the Supernatant Solution from the Precipitate

The clear solution obtained by centrifugation can be separated most simply from the precipitate by drawing off the solution with a capillary pipet. *Before* inserting the pipet in the solution, squeeze the bulb to expel air. You will find it easiest at first to draw off the solution in several portions rather than all at once. Move the tip of the pipet farther down the tube each time. If the centrifuge has an angled head, the precipitate will be banked against one side of the tube [Fig. 3.3(c)].

Some precipitates cling to the walls of the tube and will not settle further. This does no harm as long as the precipitate remains on the walls during pipetting. Other precipitates are light and float even after centrifugation. If there is any danger that they may get drawn into the pipet, wrap a wisp of cotton about the tip so that the solution is filtered through it. Very little cotton is required.

T-10. Washing Precipitates

Even after removal of the supernatant liquid, all precipitates will be wet with a small amount of the solution. The precipitate may also adsorb ions from solution to its surface. To remove these and increase the purity of the precipitate, wash it. The wash liquid is usually water, but sometimes it is advantageous to use a very dilute solution of the reagent used in the precipitation.

Washing is carried out by adding the wash liquid to the precipitate. A stirring rod is used to break up the caked precipitate and disperse it in the liquid. After centrifugation, the wash liquid may be added to the first solution if it contains an appreciable amount of ions, or it may be discarded.

As a general rule, it is more efficient to wash with two small portions of wash fluid than with one portion of the same total volume. Two washes of 10 drops each remove three times as much contaminant as one wash of 20 drops, assuming that washing is merely a matter of dilution.

T-11. Transferring Precipitates

Sometimes it is necessary to transfer a precipitate from one tube to another. The volume of solution may have been such that precipitation was carried out in two or three tubes, and the precipitates must be combined before going further. To do this, add water to one precipitate, disperse it in the water by squirting the mixture in and out with a pipet several times, then suck it up and transfer it quickly to the other tube. Clean the pipet carefully afterwards.

T-12. Testing the Acidity

Indicator papers are most useful for this purpose. Several kinds are available, ranging from familiar litmus paper to wide-range papers. The latter are impregnated with a mixture of indicators and show a succession of color changes over a wide range of pH values. Indicators can be added directly to the solution, but this is not generally desirable because they frequently become adsorbed on the surfaces of precipitates and obscure useful colors.

To use an indicator paper, dip the rod into the solution under test, withdraw it carefully, and touch it to the paper. In withdrawing the rod, keep it away from the sides of the tube, which frequently, through inadequate mixing, may be wet with acid or base. *Most of the errors committed by beginners in this test can be avoided by thorough mixing.* Test papers should never be dipped in the solution, for this soaks it up—a serious disadvantage when there are only a few drops—and contaminates the solution with indicator and paper fibers.

3.4 The Laboratory Notebook

An accurate record of experimental results and conclusions is an indispensable part of scientific work. A laboratory notebook is not a private diary but a public record. Since it is subject to scrutiny by others, it must be intelligible to anyone conversant with the subject and in such form as to leave no doubt of its honesty and reliability. In general, you should record what you did, what you saw, and what you concluded. In addition, it is a useful practice to write chemical equations for the reactions as they occur. Your instructor will specify the type of laboratory record that is appropriate for your course.

If notes are to be of any value, they should be recorded immediately. The tendency, which may be altogether subconscious, to suppress observations that do not square with a worker's preconceptions or to edit them to reinforce this bias increases after he leaves the laboratory. Notes written hours or even days after the experiment was done are unreliable as records of what actually happened.

Analysis

of Simple Substances

Is a white powder common salt or washing soda? Is the yellow pigment lead chromate or cadmium sulfide? What is the chemical in this bottle that has lost its label? These questions are easily answered after a few tests have been performed. Simple substances, because they contain one cation and one anion, present comparatively few problems to the analyst. We can confidently try to precipitate cadmium sulfide knowing that black copper sulfide cannot also form and mask the identification of yellow CdS. The vexing problems arising from the interference of one ion with the test for another are dealt with in Chapters 7–14. Here we can blithely ignore them except when cation and anion interfere with each other's identification. To be sure, we still need to keep our wits about us: the formation of an insoluble white sulfate is not sufficient evidence to conclude that barium is the cation, because there are several other cations that form such precipitates. Bias can also defeat the most careful analyst. A foolish certainty that barium is present in a salt may blind the neophyte to plain evidence that the cation is strontium.

FOUR

Detection

of the Anion

4.1 Introduction

The eleven anions that will be studied are shown in the accompanying diagram according to the position of the nonmetal in the periodic table.

IVA	VA	VIA	VIIA
$CO_3{}^{--}$	$NO_2{}^-$ $NO_3{}^-$		F^-
	$PO_4{}^{3-}$	S^{--} $SO_3{}^{--}$ $SO_4{}^{--}$	Cl^-
			Br^-
			I^-

The detection of an anion is based on the ionic reactions it undergoes, so that as we study anion analysis we learn more about proton transfer, precipitation, and redox reactions. And we become familiar with the chemistry of a number of important ions. For convenience, this chapter gives only the analytical procedures for detection of anions; the principles on which these procedures are based are covered in Chapter 1, and their specific application to the elimination tests is covered by a study guide in Chapter 5.

Anion analysis can be divided into two parts: elimination tests and identification tests. The former may indicate the absence of a group of ions. For example, if no gas is evolved when a sample is treated with dilute sulfuric acid, the absence of carbonate, sulfite, nitrite, and sulfide ions can usually be assumed. With evidence from various elimination tests at hand, it may be unnecessary to make further tests. Specific identification tests are sometimes required, and they may also be used to confirm inferences from the elimination tests.

4.2   Preparation of a Sample for Analysis

Some of the tests are carried out directly on portions of the solid sample, others on a solution of it. We are concerned here with preparation of the solution. Test the solubility of about 20 mg [T-3(b)[1]] of sample in 1 ml of distilled water. Warm if necessary in a hot water bath. If the sample is water soluble, test for the presence of "heavy metal" cations. Dissolve a little solid sodium carbonate in a few drops of water, and add a drop of this solution to the solution of the sample. If no precipitate forms, the cation is probably sodium, potassium, or ammonium. Prepare a fresh solution of the solid sample containing 20 mg/ml or 100 mg per 5 ml and use this solution directly for the tests.

If a precipitate is obtained with sodium carbonate, or if the sample is not completely soluble in water, weigh 0.10 g into a 25-ml Erlenmeyer flask of borosilicate glass (Pyrex or Kimax). Add 5 ml of distilled water, 0.50 g of anhydrous sodium carbonate, and a Carborundum chip to promote smooth boiling. Note the level of the liquid in the flask. Boil the mixture for at least 10 minutes, adding water now and then to keep the volume approximately constant. Transfer the mixture to two centrifuge tubes and settle the precipitate by centrifugation (T-8; note particularly the balancing requirement). Combine the solutions (T-9). Wash the precipitates once with water (T-10), and add the wash water to the solution. Reserve this *Prepared Solution* for the anion tests.

Combine the residues from the sodium carbonate treatment (T-11). Add 6 M acetic acid until no more gas is evolved; put in a few drops at a time and mix well after each addition (T-5). If transposition of the unknown to an insoluble carbonate was successful, the residue should dissolve in acetic acid. If transposition was incomplete, the residue will not be completely soluble in acetic acid[2]; it may contain one of the following:

$$F^-, PO_4^{3-}, S^{--}, SO_4^{--} \text{ as } BaSO_4, \text{ or } AgCl, AgBr, \text{ or } AgI.$$

[1] The reference is to Technique 3(b) in Sec. 3.3.

[2] The treatment of such residues is described in larger texts of qualitative analysis; see, for example, Chapter 13 of *Qualitative Analysis and Electrolytic Solutions* by E. J. King (Harcourt Brace Jovanovich, New York, 1959).

4.3   **Elimination Tests**

ET-1.   **Test for Strongly Acidic or Basic Anions**

The principles on which this test is based are discussed in Secs. 5.2–5.5.

Place a few drops of the solution to be tested in a semimicro test tube. Dip a clean stirring rod into the solution, withdraw it, and touch it to a short length of wide-range indicator paper. Determine the pH by comparing the color with the color standards on the indicator paper dispenser. A strongly acid reaction (pH less than 2) may be given by HSO_4^- ion or free acids. A strongly alkaline reaction (pH 10 or more) is given by S^{--}, SO_3^{--}, PO_4^{3-}, or CO_3^{--} ions. Only a *strongly* basic reaction is significant; so many ions give weakly alkaline reactions that no help is derived from such an observation. Certain salts contain the four strongly basic anions as hydrogen anions; for example, HS^-; these also give a weakly alkaline reaction.

Do not use this test on the Prepared Solution (Sec. 4.2). That solution contains excess sodium carbonate and is bound to give a strongly alkaline reaction.

ET-2.   **Test for Volatile or Unstable Acids**

The principles on which this test is based are discussed in Secs. 5.6 and 5.7.

Put sufficient solid sample in a *dry* test tube to make a layer about 1 mm deep. Add one drop of 6 M H_2SO_4 and mix by flicking the bottom of the tube with your forefinger. Watch for the evolution of gas. Note the odor by cautiously wafting some of the gas to your nose. The following ions give a reaction:

(1) Carbonate—gives colorless, odorless CO_2.
(2) Sulfite—gives colorless SO_2 with sharp odor.
(3) Nitrite—gives colorless NO, turning to red brown NO_2 in air, sharp odor.
(4) Sulfide—gives colorless H_2S with vile odor; gas turns moist lead acetate paper black.

ET-3.   **Test for Strong Reducing Substances**

The principles are discussed in Secs. 5.8–5.11.

Acidify a few drops of the solution to be tested with 6 M HCl (a tricky operation; review T-12). Prepare a fresh solution of potassium ferricyanide $(K_3Fe(CN)_6)$ by stirring a small crystal of the salt with a few drops of distilled water in a depression of the spot plate. To the acid solution under test, add one drop of $FeCl_3$ solution and one of the freshly prepared $K_3Fe(CN)_6$ solution, and let the mixture stand for no more than 5 minutes. A blue coloration or precipitate indicates the presence of a strong reducing anion: I^-, S^{--}, SO_3^{--}, or NO_2^-.

ET-4.   **Test for Strong Oxidizing Agents**

The principles are discussed in Sec. 5.12.

To 2 drops of solution in a dry test tube, add 6 drops of a saturated solution of $MnCl_2$ in 12 M HCl. Warm in a hot water bath for a minute (T-6). A yellowish brown color, which often fades to yellow as the solution cools, indicates the presence of nitrate; nitrite gives only a yellow color.

ET-5.   Test for Anions that Form Insoluble Silver Salts

The principles are discussed in Sec. 5.13.

To a few drops of solution under test in a test tube, add a drop of silver nitrate solution. If addition of a second drop produces more precipitate, continue the addition of silver nitrate until precipitation appears to be complete. After each addition mix well by flicking the bottom of the tube with your forefinger (T-4, T-5). Now add 6 M HNO_3 a drop at a time, mixing well after each drop is added, until the solution is acidic to litmus (T-12), and finally add a few drops of acid in excess. If no precipitate forms, S⁻⁻, Cl⁻, Br⁻, and I⁻ ions are absent. A black precipitate indicates S⁻⁻; the silver halides are white or pale yellow.

ET-6.   Test for Anions that Form Insoluble Calcium Salts

The principles are discussed in Sec. 5.14.

Acidify the solution to be tested with 6 M HOAc (T-12). Add several drops of calcium nitrate solution, stir, and allow to stand for 5 minutes. Then centrifuge (T-8) to bring down the precipitate, if any, so that its bulk can be observed. A precipitate at this stage will be calcium fluoride. If no precipitate forms, make the solution basic to litmus (T-12) with 15 M NH_3. If phosphate is present and the solution is basic, calcium phosphate will precipitate.

ET-7.   Test for Anions that Form Insoluble Barium Salts

The principles are discussed in Sec. 5.15.

To the solution under test, add barium chloride or nitrate solution drop by drop until precipitation is complete. Centrifuge (T-8) and draw off the solution above the precipitate (T-9) and discard it. Treat the precipitate with several drops of 6 M HCl. A white precipitate of barium sulfate, insoluble in acid, indicates the presence of sulfate or sulfite in the original sample. Sulfites give some $BaSO_4$ because they gradually become oxidized by oxygen to sulfate.

4.4   Practice on Known Samples

It is advisable to have some practical experience with the elimination tests before trying them on unknown materials. Known solutions and solid samples will be found on the side shelf. Keep a notebook record of your experiments in the form specified by your instructor. Consult Chapter 5 for help with the equations.

ET-1. Try the test on a solution of *one* of the strongly basic anions and, for comparison, on one that is not a member of that set. *Note that only the strongly basic reaction is significant.* You may find, contrary to expectations, that some anions give weakly acidic solutions (pH 5–6); this is caused by dissolved atmospheric carbon dioxide in the solutions.

ET-2. Try the test on solid salts containing the four anions. (A solution of sodium sulfide may be substituted for the solid if it is not available.)

ET-3. Try the test on all four of the strongly reducing anions.

ET-4. Try the test on solutions of nitrite and nitrate.

ET-5. Test the action of silver nitrate solution on separate portions of the standard solutions of Cl^-, Br^-, I^-, and S^{--} ions. Note carefully and record the color and appearance of the precipitates insoluble in nitric acid. For comparison, try the action of silver nitrate on the standard solution of PO_4^{3-} and test the solubility of the precipitate in nitric acid.

ET-6. Use a few drops of the standard solutions of F^- and PO_4^{3-} ions for the test. In your notebook writeup, try to distinguish the appearance of these precipitates from those of the silver halides.

ET-7. Prepare fresh solutions of SO_4^{--} and SO_3^{--} by dissolving a few crystals of their salts in a few drops of water, and try the tests on these samples. If a sulfite standard solution is available, try the test on a few drops of it too. Centrifuge and compare the bulk of the precipitates.

Prepare a summary chart for ready reference. Rule a series of vertical columns in your notebook, one for each ion, and a set of horizontal rows, one for each elimination test. Enter the *positive* results of the tests. For example, for S^{--} ion there will be entries in ET-1, ET-2, ET-3, and ET-5; in the box for ET-5 the entry would be black Ag_2S, insoluble in H_2O and dilute HNO_3.

4.5   Identification Tests

The anions are listed according to the position of their nonmetal in the periodic table, starting with the halide ions of Group VIIA and working back to the carbonate ion of Group IVA. In the absence of interfering ions the tests can be performed independently. Each test is preceded by a brief statement of the principle or chemical reaction used in it.

(a) Identification of Fluoride Ion, F^-

Hydrogen fluoride and fluorides in the presence of acid attack glass, and the surface of the freshly etched glass will shed water as if it were greasy.

$$SiO_2(s) + 4HF \rightarrow SiF_4(g) + 2H_2O$$

$$SiF_4 + 2HF \leftrightharpoons H_2SiF_6$$

In a clean, dry test tube place a few crystals of $K_2Cr_2O_7$ and tap the tube to bring them to the bottom. Add 1 ml [20 drops, T-3(a)] of 18 M H_2SO_4. Stir the mixture with a glass rod while it warms in the hot water bath (T-6). This cleaning mixture, which acts as a strong acid, dehydrating agent, and oxidizing agent, should become dark red. Withdraw the rod from the mixture now and then and examine it; when it is clean, solution will drain from it evenly in a continuous film.

Now add to the cleaning mixture about 20 mg [T-3(b)] of the solid sample. Stir it into the solution and warm the test tube in the water bath. Again withdraw the rod at frequent intervals and examine it. If the solution collects in drops along the part of the rod that was immersed in the cleaning mixture, *fluoride is present*. Continue heating and examining the rod for 5 minutes before reporting the absence of fluoride.

Cool the mixture. Empty it into a sink near the drain while a copious stream of water flows through the sink (CAUTION).

(b) Identification of Chloride Ion, Cl⁻

In the absence of interfering ions, chloride is identified by precipitation as white silver chloride, insoluble in acids, soluble in excess ammonia:

$$Cl^- + Ag^+ \rightarrow AgCl(s)$$

$$AgCl(s) + 2NH_3 \rightarrow Ag(NH_3)_2{}^+ + Cl^-$$

$$Ag(NH_3)_2{}^+ + Cl^- + 2H^+ \rightarrow AgCl(s) + 2NH_4{}^+$$

Acidify a few drops of the solution under test with 6 M nitric acid (T-12), and add a drop of AgNO$_3$ solution. A white, curdy precipitate may be AgCl. Centrifuge (T-8), and discard the supernatant liquid (T-9). Add a few drops of 6 M NH$_3$ to the precipitate. If it dissolves, acidify the solution with 6 M HNO$_3$ (T-12). The precipitate should re-form. A hazy opalescence that does not centrifuge down is caused by a trace of chloride and should not be reported.

Caution: Neophytes may mistake pale yellow AgBr or yellow AgI for AgCl. The absence of Br⁻ or I⁻ must be proved before the test for Cl⁻ can be conclusive.

(c) Identification of Bromide Ion, Br⁻

Bromide ion is oxidized to free bromine by a strong oxidizing agent such as potassium permanganate in nitric acid solution. The equation for the reaction can be derived by separating the following

$$Br^- + MnO_4{}^- \rightarrow Mn^{++} + Br_2$$

into partial equations, balancing them, and recombining (Sec. 1.12). Free bromine is more soluble in carbon tetrachloride than in water and gives an orange to brownish red color in that solvent (Sec. 2.13).

Acidify a few drops of the solution under test with 6 M HNO$_3$ and add 4 drops of acid in excess. Add 2 drops of CCl$_4$—not too much, for the color is more intense in a smaller volume. Now add 0.02 M KMnO$_4$ drop by drop, shaking after each addition, until an orange bead is obtained or the water layer remains pink for at least a minute. The *presence of bromide* is indicated by the orange lower layer.

(d) Identification of Iodide Ion, I⁻

Iodide is an active reducing agent, easily oxidized to iodine or triiodide ion by nitrous acid (Sec. 2.12); separate into partial equations and balance:

$$HNO_2 + I^- \rightarrow NO + I_2$$

Molecular iodine is much more soluble in carbon tetrachloride than in water and gives a violet solution in this nonpolar solvent (Sec. 2.13).

Acidify a few drops of the solution under test (T-12) with 6 M H$_2$SO$_4$. Add a few crystals of NaNO$_2$ and several drops of CCl$_4$. Shake vigorously to extract the iodine. The *presence of iodide* is indicated by a violet lower layer.

(e) Identification of Sulfide Ion, S⁻⁻

There is generally no difficulty in inferring the presence of sulfide from the results of the elimination tests. If a single test is required, ET-2 is the one usually employed.

(f) Identification of Sulfite Ion, SO₃⁻⁻

Sulfite ion in acidic solution is an active reducing agent and is oxidized to sulfate. It reduces permanganate, deep purple, to Mn^{++} ion, virtually colorless:

$$2H^+ + SO_3^{--} \leftrightharpoons H_2SO_3 \leftrightharpoons SO_2 + H_2O$$

$$MnO_4^- + SO_2 \rightarrow Mn^{++} + SO_4^{--} \qquad \text{(balance)}$$

The formation of sulfate is detected by the insolubility of the barium salt in acid solution:

$$Ba^{++} + SO_4^{--} \rightarrow BaSO_4(s)$$

All sulfites may be contaminated by sulfate, owing to air oxidation: $2SO_3^{--} + O_2 \rightarrow 2SO_4^{--}$. This and other sources of sulfate must be eliminated before it is possible to tell if sulfate was formed by the action of permanganate.

Acidify several drops of the solution under test with 6 M HCl (T-12). Dilute to a convenient volume by adding 10 drops of water; then add a few drops of $BaCl_2$ solution to precipitate any sulfate. Centrifuge and withdraw the solution to a clean tube (T-8, T-9). Add one drop of 0.02 M $KMnO_4$; decolorization of the permanganate and formation of white $BaSO_4$ are indications of the original *presence of sulfite*. If the chloride concentration is high, the permanganate will be slowly decolorized but no barium sulfate will form.

(g) Identification of Sulfate Ion, SO₄⁻⁻

Of all the anions that form insoluble barium salts, only sulfate ion is too weakly basic to react appreciably with acid, and thus only $BaSO_4$ will precipitate in strongly acidic solutions. If such a precipitate was obtained in ET-7 and if sulfite is proven to be absent, no other test for sulfate is required.

(h) Identification of Nitrite Ion, NO₂⁻

In acidic solution, nitrites react with sulfamic acid to give bubbles of nitrogen and sulfate ion:

$$NO_2^- + NH_2SO_3^- \xrightarrow{H^+} N_2(g) + SO_4^{--} + H_2O$$

Sulfate is detected by precipitation as barium sulfate; barium sulfamate is soluble.

To a few drops of the solution under test, add $BaCl_2$ solution. If a precipitate forms, add more $BaCl_2$ until precipitation is complete, centrifuge (T-8), and transfer the solution to a test tube (T-9). Add a few crystals of sulfamic acid to the solution, and flick the bottom of the tube with your forefinger to cause vigorous evolution of gas bubbles. The formation of both white precipitate and gas bubbles indicates the *presence of nitrite*.

(i) Identification of Nitrate Ion, NO_3^-

Nitrate ion in alkaline solution is reduced to ammonia by active metals. Aluminum or zinc may be used, but Devarda's alloy (50% Cu, 45% Al, 5% Zn) gives a smoother reaction:

$$Al + NO_3^- \rightarrow Al(OH)_4^- + NH_3(g) \qquad \text{(balance)}$$

Ammonia gas may be detected by its action on moist red litmus:

$$NH_3 + H_2O \leftrightharpoons NH_4^+ + OH^-$$

Mix several drops of the solution under test with an equal volume of 6 M NaOH. With a pipet, transfer this mixture to a dry test tube in such a way as not to wet the upper walls of the tube with the basic solution. After withdrawing the pipet, inspect the upper walls of the tube to make sure this critical requirement is satisfied. Have a piece of cotton and a strip of red litmus ready. Add a few granules of Devarda's alloy (or aluminum) and immediately push a loose cotton plug one-third of the way down the tube to filter out spray. Warm briefly in the water bath to induce a vigorous reaction. Withdraw the tube and insert a piece of red litmus, first bending the strip into a V and moistening the fold. The dry upper ends will support the litmus so that is does not reach the cotton (Fig. 4.1). Allow the tube to stand for several minutes. The *presence of nitrate* is indicated if the bottom tip of the litmus turns uniformly blue.

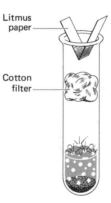

Litmus paper

Cotton filter

Fig. 4.1. The nitrate test.

Nitrite ion will form ammonia under these conditions, and ammonium ion will also. If the cation of the simple substance is known to be ammonium ion it must be removed before making the test. To the sample in a casserole, add 10 drops of 6 M NaOH and evaporate to small volume over a microflame (T-7):
$$NH_4^+ + OH^- \rightarrow NH_3(g) + H_2O.$$

(j) Identification of Phosphate Ion, PO_4^{3-}

When an excess of ammonium molybdate is added to a solution of a phosphate that is 5–10% HNO_3 by volume, bright yellow ammonium molybdophosphate precipitates. The composition of the precipitate is approximated by $(NH_4)_3[PMo_{12}O_{40}] \cdot 6H_2O$. The radical in brackets is the anion of an *hetero-polyacid*: *hetero-* because it contains two kinds of atoms other than oxygen and hydrogen ($H_6Mo_7O_{24}$, by contrast, is an *isopolyacid*); *poly-* because it contains more than one molybdenum atom. The ion has a cagelike structure.

Acidify a few drops of the solution under test with $6\,M\,HNO_3$, and add 2 drops of acid in excess. Warm the solution for a minute in a hot water bath; it should reach a temperature of 40–50° (not too hot to hold) for fast reaction in the next step. Withdraw the tube from the water bath, and add 2 drops of ammonium molybdate reagent $[(NH_4)_2MoO_4]$. Allow to stand for 5 minutes. A bright yellow precipitate of ammonium molybdophosphate indicates the *presence of phosphate*. White MoO_3 may form if the phosphate is absent or if the solution is too hot.

(k) Identification of Carbonate Ion, CO_3^{--}

The test for carbonates is a modification of ET-2:

$$2H^+ + CO_3^{--} \rightarrow CO_2(g) + H_2O$$

When the gas is absorbed in barium hydroxide solution, a white precipitate of $BaCO_3$ is obtained:

$$Ba^{++} + 2OH^- + CO_2 \rightarrow BaCO_3(s) + H_2O$$

The result will be conclusive only if sulfite is known to be absent (the elimination tests will show this), for sulfur dioxide is also liberated by acid and also gives a white turbidity with barium hydroxide solution.

Add 1 or 2 drops of $6\,M\,H_2SO_4$ to 20–30 mg of solid sample (*not* the Prepared

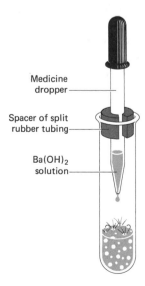

Medicine
dropper

Spacer of split
rubber tubing

Ba(OH)₂
solution

Fig. 4.2. The carbonate test.

Solution). Have a medicine dropper (not a capillary pipet) ready with a drop of $Ba(OH)_2$ solution in the tip. After adding the acid, quickly insert the dropper without squeezing the bulb (Fig. 4.2). The *presence of carbonate* is indicated by the formation of a white turbidity.

4.6 Analysis of a Soluble Salt

As a first "unknown" for anion analysis, a sodium, potassium, or ammonium salt is suggested. It will be water soluble and will contain one of the anions studied in this chapter. Tests for the three cations are given in Sec. 6.5. Prepare a solution of 0.10 g in 5 ml of distilled water for the anion analysis. A portion of this acidified with acetic acid can be used for the sodium and potassium tests.

Look for strong, definite tests; distrust—but do not necessarily ignore—faint color changes and slight turbidities. If you have any doubt about the validity of a test on the unknown, compare it with the same test on an authentic solution of the ion. Beware of preconceptions! Through faulty technique or mere inattention, you might fail to note evidence for sulfite in the elimination tests. If you think sulfite is absent and test for carbonate, you will get a false test. In other words, even though your salt contains only one anion, do not ignore the possibility of interference or a false test by an ion other than the one you think may be present.

Your instructor will specify the type of laboratory record he expects you to make of the analysis.

4.7 Analysis of Other Salts

Other simple salt unknowns may contain one of the anions studied in this chapter and one of the cations listed in Sec. 6.1. Examine the original sample as directed in Sec. 6.2. Then identify the anion. To get the anion into solution it may be necessary to transpose the sample with sodium carbonate as directed in Sec. 4.2. The cation analysis begins with preparation of a solution as described in Sec. 6.3. Then do the preliminary tests in Sec. 6.4. Finally, perform the identification test, if necessary, for the suspected cation (Sec. 6.5).

FIVE

Principles

of Anion Analysis

5.1 Introduction

The anion elimination· tests involve types of reactions that were discussed in Chapter 1: proton transfer, redox, and precipitation. The present chapter re-examines such reactions in a format designed for self-study. Accordingly, there are frequent blanks to be filled in or a choice of words to be made to fit the context. The text is separated into frames, short sections that contain one or more blanks or word choices. The answers for one frame are given in the right-hand part of the next one so that you can check yourself. For most effective use of these study guides it is recommended that you

(1) write the answers rather than just say them to yourself,

(2) keep the given answers covered until you have written, and

(3) limit such study sessions to a half-hour or so, rather than attempting to do the entire chapter in one sitting.

To facilitate studying brief portions at a time, the chapter is divided into 16 sections, which are grouped under three main heads according to the particular type of reaction under study. Review exercises are given at the end of each of these groups of sections.

PROTON TRANSFER REACTIONS

5.2 Acids and Bases

Acids have certain characteristics in common—for example, sour taste, effect on indicators, ability to neutralize bases—that we use to recognize an acid in the laboratory. In doing so, we make no hypothesis about the structure of the acid. In fact, various structural conceptions of an acid are consistent with these characteristics. The most useful is that of Brønsted:

Acids are proton donors.

Protons are hydrogen nuclei: we represent them by p to distinguish them from hydrated protons in solution which we shall subsequently represent by H^+. The definition of acid is then illustrated by

Acetic acid: $CH_3COOH \rightarrow p + CH_3COO^-$

1. _____ acid: $HNO_2 \rightarrow p +$ _____	
2. Hypochlorous acid: $HOCl \rightarrow p +$ _____	Nitrous; NO_2^-
3. _____ ion: $NH_4^+ \rightarrow p + NH_3$	ClO^- (or OCl^-)
4. All such ionization reactions are reversible: the ion or molecule on the right-hand side of the equation can take back the proton and form the acid on the left-hand side. Brønsted therefore complemented his definition of acid by defining a *base* as a proton _____.	Ammonium
5. An acid and a base that differ only by a proton _____ $\leftrightharpoons$ base + p are called a *conjugate* pair.	acceptor
6. The base conjugate to CH_3COOH is _____.	Acid
7. The acid conjugate to NH_3 is _____.	CH_3COO^-

8. Water, an *ampholyte*, has both acidic and basic properties. As an acid it donates a proton $$H_2O \rightarrow p + \underline{\quad\quad}$$ and as a base it accepts one $$p + H_2O \rightarrow \underline{\quad\quad}$$	NH_4^+
9. Neither property is very well developed; water is a very strong/weak acid or base. The concentration of hydroxide ion in pure water is only 10^{-7} M at 25°, and the concentration of hydrated protons is likewise only 10^{-7} M, making the pH of pure water equal to 7.	OH^- H_3O^+
10. For a neutral solution, then, $[H_3O^+] = [OH^-]$. For an acid solution, $[H_3O^+] > [OH^-]$ and the pH is greater/less than 7 at 25°.	weak
11. For a basic solution, $[H_3O^+]$ _____ $[OH^-]$ and the pH is greater than 7.	less
	<

5.3   Reactions of Univalent Anions with Water (ET-1)

12. Consider the salt sodium acetate, $NaCH_3COO$ or NaOAc for simplicity. Most salts are strong/weak electrolytes. Sodium acetate is not an exception to this rule, so we classify it as a strong/weak electrolyte.	
13. On this basis, when sodium acetate dissolves in water we expect it to break up completely into Na^+ and _____ ions.	strong strong
14. The rule is that most acids are strong/weak electrolytes. Acetic acid is/is not one of the exceptions, so acetic acid is a weak electrolyte.	OAc^-
15. We expect a small fraction of all the acetic acid molecules to lose protons and form _____ ions.	weak is not

16. Conversely, acetate ions can function as <u>acids/bases</u>, accepting protons to give molecular acetic acid.	acetate
17. Acetate ions, which are basic, and water molecules, which are weakly acidic (that is, capable of donating _____) should be able to react by proton transfer: $$OAc^- + H_2O \leftrightharpoons HOAc + \underline{\quad}$$	bases
18. Because water is a <u>good/poor</u> proton donor, only one acetate in 10,000 undergoes this reaction in 0.1 M sodium acetate solution.	protons OH$^-$
19. We can nevertheless detect this slight degree of reaction by its considerable effect on the concentration of hydroxide ion. The hydroxide ion concentration of water is _____ and changes to 10^{-5} M when sodium acetate is added.	poor
20. This increase in hydroxide ion concentration is balanced by a decrease in hydrogen ion concentration from _____ to 10^{-9} M, equivalent to a pH change from 7 to _____.	10^{-7} M
21. The product of hydrogen and hydroxyl ion concentrations remains constant, equaling K_w, the ion product of water: $$10^{-7} \times 10^{-7} = 10^{-5} \times 10^{-9} = \underline{\qquad}$$	10^{-7} M $9 \, (-\log 10^{-9})$
22. A sodium acetate solution is basic because [OH$^-$] is <u>greater/less</u> than the hydrogen ion concentration.	10^{-14}
23. The reaction we have described is often called the *hydrolysis* of sodium acetate. In general, salts of weak acids hydrolyze to give <u>acidic/basic</u> solutions.	greater
24. In Brønsted terminology, such salts have anions that act as <u>acids/bases</u> toward water acting as an <u>acid/base</u>.	basic

25. The sodium ion of the salt sodium acetate has been omitted from consideration and from the equation in frame 17 because it does not react with water. Sodium ion of the salt, which is almost inert, must be distinguished from the atom of sodium metal, which is very reactive: $$Na^+ + H_2O \rightarrow \text{ no reaction}$$ $$2Na + 2H_2O \rightarrow \underline{\hspace{1cm}} + 2Na^+ + 2OH^-$$	bases acid
26. A sodium ion has one less electron/proton than a sodium atom. They are chemically very distinct entities.	$H_2(g)$
27. Most acids, like acetic acid, are strong/weak. Hydrochloric acid, one of the exceptions to the rule, is strong/weak.	electron
28. In dilute aqueous solutions, hydrochloric acid is completely ionized. If the reaction $$H_2O + HCl \rightarrow Cl^- + \underline{\hspace{1cm}}$$ is complete, the reverse reaction $$H_3O^+ + Cl^- \rightarrow \underline{\hspace{1cm}} + H_2O$$ must not go at all.	weak strong
29. In dilute solution, chloride ion has no appreciable basic properties. Another way of saying this is that chloride ion is/is not hydrolyzed.	H_3O^+ HCl
30. Basicity constants, K_b, measure the strengths of bases relative to that of OH^- taken as unity. We usually derive them from acidity constants K_a by the relation $$K_b = K_w/K_a$$ where K_w is the $\underline{\hspace{2cm}}$ of water, with the value 10^{-14} at 25°C.	is not
31. The acid conjugate to Cl^- is $\underline{\hspace{1cm}}$ with an estimated acidity constant of 10^{+6}, so K_b of Cl^- is $10^{-14}/10^6 = \underline{\hspace{1cm}}$, a negligibly small value.	ion product

32. The acid conjugate to OAc^- is _____ with an acidity constant of 1.8×10^{-5}. The basicity constant of acetate ion is therefore equal to _____$/(1.8 \times 10^{-5})$, or 5.6×10^{-10}.	HCl 10^{-20}
33. The relative basic strengths of OH^-, OAc^-, and Cl^- are therefore $1 : 5.6 \times 10^{-10} : $_____.	HOAc 10^{-14}
34. Suppose we dissolve the salt potassium chloride in water. Like most salts, it is a strong/weak electrolyte and breaks up completely/incompletely into K^+ and Cl^- ions in water.	10^{-20}
35. Like Na^+ ion, the K^+ ion is rather inert chemically and does/does not react with water. In view of the fact that Cl^- ion is a very strong/weak base, we expect that it will/will not react appreciably with water in dilute solutions.	strong completely
36. Potassium chloride is an example of a salt that does not hydrolyze. Its cation is not acidic and its anion is not _____.	does not weak will not
37. Potassium chloride solutions are therefore acidic/basic/neutral and should have a pH of 7. (They are in fact usually weakly acidic because they absorb carbon dioxide from air.)	basic
	neutral

5.4 Reactions of Bivalent and Tervalent Anions with Water (ET-1)

38. Although Cl^- ion is not appreciably basic, we might expect different behavior from a doubly charged anion like S^{--}. Sulfide and chloride ions are about the same size but differ in charge. If we place the two ions the same distance from a proton, the attraction for the positive charge by S^{--} will be _____ times as great as that by Cl^-.	

39. We therefore expect S^{--} to be less/more basic than Cl^-.	two
40. Sulfide salts, for that reason, are extensively hydrolyzed: $$S^{--} + H_2O \leftrightharpoons HS^- + \underline{\hspace{1cm}}$$	more
41. If S^{--} ion is a stronger base than Cl^- ion, the conjugate acids must show the reverse relationship. The acid conjugate to S^{--} is _____. Thus, HCl is a stronger/weaker acid than HS^-.	OH^-
42. The acidity constant of HS^- is about 10^{-14}, so that the basicity constant of S^{--} must be $10^{-14}/10^{-14}$ or _____, making S^{--} ion as strong a base as OH^- ion.	HS^- stronger
43. Sulfide ion must likewise be a stronger/weaker base than acetate ion, and HS^- ion must be a stronger/weaker acid than HOAc.	1
44. Comparing 0.1 M solutions of Na_2S and NaOAc, we expect the degree of hydrolysis of Na_2S to be larger/smaller than that of NaOAc.	stronger weaker
45. Bearing in mind that hydrolysis of either of these salts produces a surplus of hydroxide ions over hydrated protons, we nevertheless expect the surplus to be different in the two solutions because of the different degrees of hydrolysis. The hydroxide ion concentration of 0.1 M Na_2S should be greater/smaller than that of 0.1 M NaOAc.	larger
46. The hydrogen ion concentration of the sodium sulfide solution is therefore greater/smaller than that of the sodium acetate solution, and the pH of the sulfide solution is larger/smaller.	greater
	smaller larger

To summarize:

> The more basic the anion of the salt,
>   the weaker is its conjugate acid,
>   the greater is its degree of hydrolysis,
>   and the larger is the pH of its solution.

---

47. It may have been noticed that in the reaction of sulfide ion with water, only one water molecule was used and only one proton was transferred:

$$S^{--} + H_2O \rightleftharpoons HS^- + OH^- \qquad (1)$$

A second step is also possible:

$$HS^- + H_2O \rightleftharpoons H_2S + \underline{\qquad} \qquad (2)$$

---

48. We may expect the second step to be less important than the first for two reasons.

    (a) The basic anion HS^- is only singly charged, making its attraction for a proton <u>less</u>/more than that of doubly charged S^{--} ion.

    (b) Step (2) occurs subsequent to step (1) and therefore takes place in a solution that already contains a rather high concentration of hydroxide ion. By the common-ion effect, this tends to repress any further reaction that produces more _____ ion.

OH^-

---

49. We call the reaction represented by equation (1) the *principal hydrolysis reaction* of sodium sulfide. For many purposes we can neglect the second step, which occurs to a <u>larger</u>/smaller extent.

less
hydroxide

---

50. Now we should be able to use these principles to make predictions about other anions. Consider the ions Br^-, PO_4^{3-}, and F^-. The conjugate acids are HBr, _____, and HF.

smaller

---

51. Of these conjugate acids, HBr is <u>strong</u>/weak; HF is a moderately weak acid with an acidity constant of 6.7×10^{-4}; and HPO_4^{--} ion is a very weak acid with an acidity constant of 4.2×10^{-13}.

HPO_4^{--}

52. The basicity constant of F^- ion is equal to

$$K_b(F^-) = \frac{K_w}{K_a(HF)} = \frac{10^{-14}}{\underline{\hspace{2cm}}} = 1.5 \times 10^{-11}$$

| | strong |

53. The basicity constant of PO_4^{3-} ion is equal to

$$K_b(PO_4^{3-}) = \frac{\underline{\hspace{2cm}}}{4.2 \times 10^{-13}} = 2.4 \times 10^{-2}$$

| | 6.7×10^{-4} |

54. The most basic of the three anions is _____, and the least basic is _____.

| | 10^{-14} |

55. If we compare 0.1 M solutions of NaBr, Na_3PO_4, and NaF, the degree of hydrolysis should be least for _____ and greatest for _____.

| | PO_4^{3-}
 Br^- |

56. The pH will be greatest for 0.1 M _____ and least for 0.1 M NaBr.

| | NaBr
 Na_3PO_4 |

| | Na_3PO_4 |

5.5   Reactions of Ampholytic Anions with Water (ET-1)

57. A 0.1 M solution of potassium hydrogen sulfate is acidic, not basic, so its pH is greater/less than 7.

58. How do we reconcile this observation with the preceding discussion? Potassium hydrogen sulfate (formerly called potassium bisulfate) is a salt and should be a strong/weak electrolyte, completely broken up into HSO_4^- and _____ ions.

| | less |

59. As before, we regard an alkali metal ion like K^+ as virtually inert, so that the behavior of _____ ion must be responsible for the acid solution.

| | strong
 K^+ |

60. The hydrogen sulfate ion can, in principle, be considered to be an ampholyte. In other words, it could act as an acid and also as a _____.

| | HSO_4^- |

61. The acidic behavior is represented by $$HSO_4^- + H_2O \leftrightharpoons H_3O^+ + \underline{} \qquad (1)$$	base
62. The basic behavior is represented by $$HSO_4^- + H_2O \leftrightharpoons \underline{} + H_2SO_4 \qquad (2)$$	SO_4^{--}
63. But in dilute solutions H_2SO_4 is known to lose one proton completely; with respect to this change it is a strong/weak acid like HCl.	OH^-
64. So HSO_4^-, like Cl^-, must have no appreciable _____ character in dilute aqueous solution.	strong
65. The extent of the basic reaction [equation (2), frame 62] should therefore be virtually _____.	basic
66. On the other hand, HSO_4^- is a rather good acid (acidity constant about 10^{-2}). Reaction (1) will therefore occur to a considerable extent, putting into solution a surplus of hydrated hydrogen ions, with the result that $[H_3O^+]$ _____ $[OH^-]$.	nil (zero)
67. This relationship is characteristic of acid solutions, pH _____ 7.	>
68. For many other ampholytic anions, the acidic and basic functions are more closely balanced. Some, such as $H_2PO_4^-$ and HSO_3^- ions, give an acidic reaction, but never as acidic as equimolar HSO_4^-. Others, such as HPO_4^{--} or HCO_3^-, give a basic solution, but never as basic as PO_4^{3-} or CO_3^{--}. 　　To take HCO_3^- as an illustration, the acid reaction is $$HCO_3^- + H_2O \leftrightharpoons H_3O^+ + \underline{}$$ with an acidity constant $$K_2(H_2CO_3) = K_a(HCO_3^-) = 4.7 \times 10^{-11}$$	<
69. The basic reaction is $$HCO_3^- + H_2O \leftrightharpoons OH^- + \underline{}$$ with a basicity constant $$K_b(HCO_3^-) = K_w/K_1(H_2CO_3) = 2.3 \times 10^{-8}$$	CO_3^{--}
70. The basicity constant of HCO_3^- ion is therefore larger/smaller than its acidity constant.	H_2CO_3

71. On that basis, HCO_3^- ion is stronger/weaker as a base than it is as an acid. Its solution is weakly basic, pH about 8.	larger
72. A final point about notation: up to now we have written H_3O^+ for the hydrated proton. There is, in fact, good evidence for further hydration of this ion with three water molecules to give $H_9O_4^+$. Virtually all other ions are also hydrated; yet we do not customarily show this. Hereafter, in the interests of simplicity and consistency, we shall adopt the practice of writing H^+ as shorthand for the hydrated proton. In place of $$HSO_4^- + H_2O \leftrightharpoons \text{_____} + SO_4^{--}$$ we shall write $$HSO_4^- \leftrightharpoons H^+ + SO_4^{--}$$ with the understanding that all the symbols stand for hydrated species and that water plays a vital, if implicit, role in the reaction.	stronger
	H_3O^+

5.6 Formation of Unstable or Volatile Acids. Proton Transfer (ET-2)

73. Hydrogen sulfide is only sparingly soluble in water. A saturated solution is 0.1 M in H_2S at 25° and 1 atm. Lower concentrations of hydrogen sulfide are produced by hydrolysis of sulfide salts; these solutions are super-saturated/unsaturated.	
74. As we saw in Sec. 5.4, the hydrolysis reactions are Principal: $S^{--} + H_2O \leftrightharpoons OH^- + $ _____ Secondary: $HS^- + H_2O \leftrightharpoons H_2S + $ _____	unsaturated
75. In these reactions water acts as a very weak proton acceptor/donor. To get greater conversion of S^{--} to H_2S, so that the solubility will be exceeded and H_2S gas will be evolved, we need to use a stronger/weaker proton donor than water.	HS^- OH^-

76. The obvious type of substance to try is a strong <u>acid/base</u>, such as $HClO_4$, $HClO_3$, HI, HBr, or _____ .	donor stronger
77. The acid actually used in ET-2 is dilute _____ . It is easy to come by, gives off no fumes of its own (HCl, HBr, HI do), is not a good oxidizing agent (HNO_3 and $HClO_3$ are), and its anion is not a reducing agent (I^- is).	acid H_2SO_4 (or HCl or HNO_3)
78. Concentrated sulfuric acid is largely molecular H_2SO_4. The dilute acid is approximately two-thirds HSO_4^- and one-third SO_4^{--}, along with sufficient _____ ions to make the solution electrically neutral.	H_2SO_4
79. Its reaction with sulfide can involve either $$2H^+ + S^{--} \rightarrow \text{_____}$$ or $$2HSO_4^- + S^{--} \rightarrow H_2S(g) + \text{_____}$$ For simplicity we shall use only the first.	H^+
80. Such a reaction with a strong proton donor produces a concentration of molecular hydrogen sulfide that is sufficient to saturate the solution, allowing excess to escape as a _____ . Hydrogen sulfide is a _____ acid.	$H_2S(g)$ $2SO_4^{--}$
81. The reaction of sulfuric acid with a carbonate or a sulfite likewise produces a gas. Initially the bases are protonated: $$2H^+ + CO_3^{--} \rightarrow H_2CO_3$$ $$2H^+ + SO_3^{--} \rightarrow \text{_____}$$	gas volatile
82. But the two acids are unstable, breaking down into water and an acidic oxide: $$H_2CO_3 \rightarrow \text{_____} + CO_2(g)$$ $$H_2SO_3 \rightarrow \text{_____} + H_2O$$	H_2SO_3
83. Addition of sulfuric acid to carbonates or sulfites thus results in evolution of the gaseous oxides CO_2 and _____ .	H_2O $SO_2(g)$

84. By combining the two stages (frames 81 and 82) we can show evolution of gas in single equations: $$2H^+ + CO_3^{--} \rightarrow H_2O + \underline{\hspace{1.5cm}}$$ $$2H^+ + SO_3^{--} \rightarrow \underline{\hspace{1.5cm}} + SO_2(g)$$	SO_2
85. Nitrous acid, HNO_2, is a <u>strong/weak</u> acid, so when dilute sulfuric acid is added to a nitrite, formation of the molecular acid occurs $$H^+ + NO_2^- \rightarrow \underline{\hspace{1.5cm}}$$	$CO_2(g)$ H_2O
The subsequent decomposition of nitrous acid is more complex than that of the other acids and will be dealt with in Sec. 5.7.	weak HNO_2

REVIEW EXERCISES

Test the understanding you acquired in working through the preceding sections of the study guide on proton transfer reactions by answering the following questions.

5.1. What is the acid conjugate to CO_3^{--}? To HCO_3^- ion? What is the base conjugate to $H_2PO_4^-$ ion? Why is the last of these ions called an ampholyte?

5.2. We have seen that the chloride ion has virtually no basic properties in dilute solution. Name some other anions of the same type. We have seen that sodium and potassium ions are inert and do not react with water. Name another such cation.

5.3. The pH of 0.1 M KCN is found by experiment to be greater than that of 0.1 M NaSCN. Which of the following statements are true?
(a) KCN is more highly hydrolyzed than NaSCN.
(b) KCN is more highly ionized than NaSCN.
(c) HCN is a stronger acid than HSCN.
(d) KOH is a stronger base than NaOH.
(e) The basicity constant of CN^- is larger than that of SCN^-.

5.4. Criticize the following equations:
(a) $KI + H_2O \leftrightharpoons KOH + HI$
(b) $CO_3^{--} + H_2O \leftrightharpoons 2OH^- + CO_2(g)$

5.5. The acidity constant for loss of one proton by H_2S is 1.0×10^{-7}; the acidity constant of HS^- ion is about 1×10^{-14}. What is the value of the basicity constant of HS^- ion? Is HS^- stronger as an acid or as a base? Will the pH of 0.1 M NaHS be greater than, equal to, or less than 7?

REDOX REACTIONS

5.7   Formation of Unstable or Volatile Acids. Redox (ET-2)

1. When dilute sulfuric acid is added cautiously to a cold, dilute solution of a nitrite, a blue solution of weakly ionized HNO_2, _____ acid, is formed.	
2. At room temperature HNO_2, like H_2CO_3 and _____, is unstable.	nitrous
3. Carbonic acid and sulfurous acid break down into water and an oxide. An acid and an oxide that differ only by H_2O are said to be *conjugate*. Thus, H_2CO_3 is conjugate to the oxide _____.	H_2SO_3
4. Oxidation numbers of elements in compounds like H_2CO_3 and CO_2 are assigned on the basis of certain conventions. The oxidation number of oxygen in such compounds is usually taken to be _____ and that of hydrogen to be _____.	CO_2
5. The oxidation number of C in CO_2 must be $4+$ since the two oxygen atoms contribute $4-$ and the sum of oxidation numbers for a molecule must be _____.	$2-$ $1+$
6. By the same reasoning, the oxidation number of C in H_2CO_3 must also be $4+$ because two hydrogen atoms contribute $2+$ and three oxygen atoms contribute _____.	zero
7. Thus, both in H_2CO_3 and in its conjugate oxide CO_2, the oxidation number of the nonmetal carbon is _____.	$6-$
8. The nonmetal, as a general rule, has the same oxidation number in conjugate _____ and oxide.	$4+$ (the same)
9. The oxidation number of nitrogen in HNO_2 is _____.	acid
10. The oxide conjugate to HNO_2 must also contain nitrogen in an oxidation state of _____. This oxide is N_2O_3, an unstable one.	$3+$ $(1+x-4=0)$

11. The stable oxides of nitrogen are
    Nitrous oxide, N_2O, with oxidation number of
    nitrogen = _____
    Nitric oxide, NO, with oxidation number of nitrogen
    = _____
    Nitrogen dioxide, NO_2 with oxidation number of
    nitrogen = _____
	$3+$

12. Decomposition of nitrous acid gives, not unstable N_2O_3,
    but the more stable oxide NO and also nitric acid. The
    reaction is represented by
 	
    $$3HNO_2 \leftrightharpoons H^+ + NO_3^- + 2NO(g) + H_2O \quad (1)$$
 	
    In this reaction the oxidation number of one nitrogen
    atom increases from $3+$ in HNO_2 to _____ in NO_3^-.
	$1+$
	$2+$
	$4+$

13. Because the oxidation number of this nitrogen atom has decreased/increased, it is said to be *oxidized*.	$5+$ $(x-6=-1)$

14. In reaction (1) (frame 12), the other two nitrogen atoms in
    two HNO_2 molecules undergo *reduction*. The oxidation
    number of each of these atoms goes from $3+$ in HNO_2 to
    _____ in NO.
	increased

15. Equation (1) (frame 12) represents an oxidation–
    reduction or *redox* reaction. Note that the amount of
    oxidation and reduction, measured in units of oxidation
    number change, must match:
    One N atom goes from $3+$ to $5+$, an increase of
    _____ units, and
    Two N atoms go from $3+$ to $2+$, a total decrease of
    _____ units.
	$2+$

16. Since the same substance, HNO_2, is undergoing both
    oxidation and _____, reaction (1) is often
    referred to as an *auto-oxidation reduction* reaction. (Auto-
    comes from Greek *autos*, meaning self.)
	2
	2

17. Let us verify that no oxidation–reduction occurs in the
    reaction $H_2SO_3 \rightarrow H_2O + SO_2$. The oxidation number of
    sulfur in H_2SO_3 is _____, and in SO_2 it is also _____,
    no change.
	reduction

18. When nitric oxide, _____, is exposed to the oxygen of air, it becomes converted to red brown NO_2, _____ _____.	4+ 4+
19. A balanced equation for this reaction is $\qquad$ $NO + O_2 \rightarrow$ $\qquad$ NO_2 $\qquad$ (2)	NO nitrogen dioxide
20. In this reaction the oxidation number of oxygen changes from _____ in O_2 to _____ in NO_2. Oxygen is oxidized/reduced.	2 2
21. The oxidation number of nitrogen changes from _____ in NO to _____ in NO_2. Nitrogen atoms (and, by extension, NO itself) are said to be oxidized.	0 2− reduced
22. The reactant or reagent causing oxidation of something else is called the oxidant or *oxidizing agent*. In reaction (2) (frame 19), the oxidizing agent is _____.	2+ 4+
23. The oxidizing agent is itself reduced, and the reagent that is responsible for this is called the *reducing agent* or reductant. Since oxidation and reduction go together, the reducing agent must be oxidized. It is the nitrogen part of NO that undergoes oxidation, but the reagent—what we take off the shelf and add to the other substance—is not nitrogen atoms but a complete compound. So in reaction (2) (frame 19) the reducing agent is _____ oxide.	O_2
	nitric

5.8 Test for Strong Reducing Substances. The Reduction Half-Reaction (ET-3)

24. In this test the presence of strong reducing agents results in the formation of a color or precipitate of Prussian blue, $KFeFe(CN)_6$. The formula is written with FeFe, rather than Fe_2, to emphasize that this complex salt contains two kinds of iron atoms. The common oxidation states of iron are 0, 2+, and 3+. Metallic or uncombined iron has an oxidation number of _____.	

25. The oxidation number of iron in Fe^{++} ion, $FeCl_2$, $FeSO_4$, and similar compounds is _____. The compounds used to be called ferrous salts, and Fe^{++} was ferrous ion.

zero

26. Since the suffix -ous corresponded to an oxidation number of $2+$ in the ions of some elements, $1+$ in those of other elements, and $3+$ in still others, confusion frequently resulted. We have therefore taken more recently to indicating the oxidation number with a Roman numeral. The name of $FeCl_2$ is now iron(II) chloride, more explicit if less mellifluous than ferrous chloride. Likewise, $FeSO_4$ is named iron(II) sulfate and FeS is _____ _____.

$2+$

27. Cyanide ions (CN^-) and iron(II) ions (Fe^{++}) combine to form the complex $Fe(CN)_6^{4-}$ ion. Note that the charge on the complex is the sum of the charges on the iron ion ($2+$) and six cyanide ions (_____), a net value of $4-$.

The systematic name of this complex is hexacyanoferrate(II) ion, but it is more commonly called ferrocyanide ion.

iron(II) sulfide

28. Iron in an oxidation state of $3+$ is found in Fe^{3+}, $FeCl_3$, $Fe(NO_3)_3$, $NH_4Fe(SO_4)_2 \cdot 12H_2O$ (ferric alum), and similar compounds. These used to be called ferric salts. Now, according to the systematic rule, $FeCl_3$ is named iron(III) chloride and $Fe(NO_3)_3$ is _____ _____.

$6-$

29. Iron(III) ion also forms a complex by combining with six cyanide ions. The charge on the complex is $(3+)+(6-)=$ _____, and its formula is therefore _____.
It is called the hexacyanoferrate(III) ion or ferricyanide ion.

iron(III) nitrate

30. Prussian blue contains one iron(III) ion and the ferrocyanide ion: $KFe(III)Fe(II)(CN)_6$. Like many compounds that contain the same metal in two oxidation states, it is deeply colored. This dark blue substance was one of the first synthetic pigments, having been made originally in the eighteenth century. It can be prepared most directly by mixing an iron(III) salt with potassium ferrocyanide:

$$K^+ + Fe^{3+} + \underline{\hspace{3cm}} \rightarrow KFeFe(CN)_6$$

$3-$
$Fe(CN)_6^{3-}$

31. In ET-3 we mix iron(III) chloride, potassium ferricyanide, and an acid solution of the substance under examination. This combination can give Prussian blue only if some ferricyanide is converted to ferrocyanide. Since this involves reduction of iron from an oxidation state of _____ in $Fe(CN)_6^{3-}$ to _____ in $Fe(CN)_6^{4-}$, we need a reducing agent.	$Fe(CN)_6^{4-}$
32. The formation of Prussian blue is therefore visual evidence of the presence of a _____ agent in the reaction mixture.	3+ 2+
33. The change in ionic charge of the cyanide complex from 3− to 4− can be accomplished by adding one unit of negative charge, namely, one _____.	reducing
34. We can now represent the reduction part of the overall reaction in ET-3 by a partial equation $Fe(CN)_6^{3-} + e^- \rightarrow$ _____	electron
35. For conciseness we can combine the reduction with formation of Prussian blue (frames 30 and 34) $K^+ + Fe^{3+} + Fe(CN)_6^{3-} + e^- \rightarrow$ _____	$Fe(CN)_6^{4-}$
36. Because reduction is always accompanied by oxidation, there will necessarily be a second partial equation for that process. Every electron gained in reduction must be matched by one _____ in oxidation.	$KFeFe(CN)_6$
	lost

A firm command of the following basic concepts is required if you are to follow the subsequent discussion.

> *Reduction*
>   corresponds to a decrease in oxidation number
>   or a gain of electrons.
> *Oxidation*
>   corresponds to an increase in oxidation number
>   or a loss of electrons.
> An *oxidizing agent*
>   causes oxidation of something else, undergoes
>   reduction itself, and gains electrons.
> A *reducing agent*
>   causes reduction of something else, undergoes
>   oxidation itself, and loses electrons.

5.9   Test for Strong Reducing Substances. The Oxidation Half-Reaction (ET-3)

37. Having seen that we need a reducing agent to get Prussian blue from ferricyanide, let us examine the list of anions we are considering for possible reducing agents. To be reducing agents they must be capable of being oxidized. That requirement immediately rules out several anions. In sulfate ion, SO_4^{--}, the oxidation number of sulfur is _____.	
38. The maximum oxidation number of a nonmetal is given by the group it is in in the periodic table, for example, $4+$ for C, a member of Group IV; $7+$ for Cl, a member of Group VII. Sulfur is a member of Group VI, so its maximum oxidation number is _____.	$6+$ $(x-8=-2)$
39. The oxidation number of S in SO_4^{--} is already at its maximum value, so it cannot be oxidized further. Thus, sulfate ion cannot be a _____ agent. (Conceivably, the O^{--} in SO_4^{--} might be oxidized to O_2, but this does not happen.)	$6+$
40. Nitrogen and phosphorus are members of Group _____ in the periodic table, so that their maximum oxidation number is $5+$.	reducing
41. The oxidation number of N in NO_3^- is _____ and that of P in PO_4^{3-} is _____.	V
42. Nitrate and phosphate ions cannot be reducing agents because their nonmetals are/are not in their highest oxidation states.	$5+$ $5+$
43. Carbon is a member of Group _____ in the periodic table and its oxidation number in carbonate ion is _____, so that carbonate can/cannot be a reducing agent.	are
44. The other anions under consideration, F^-, Cl^-, Br^-, I^-, S^{--}, SO_3^{--}, and NO_2^-, contain a nonmetal in an oxidation state below its maximum value and are, in principle, capable of being oxidized/reduced.	IV $4+$ cannot

45. The halide ions can be converted to the free halogen; for example, $$2F^- \rightarrow F_2 + \underline{\hspace{2cm}}$$	oxidized
46. Loss of $2\,e^-$ per $2\,F^-$ is equivalent to loss of $\underline{\hspace{1cm}}$ electron per fluoride or an increase in oxidation number of fluoride from $\underline{\hspace{1cm}}$ to zero.	$2e^-$
47. The ease with which a halide ion loses an electron depends on its size. Fluoride ion, the smallest, holds its electron very tightly and is difficult/easy to oxidize.	one $1-$
48. Conversely, the reaction $2e^- + F_2 \rightarrow 2F^-$ is very easy to accomplish. Fluorine is one of the best of all oxidizing agents, but fluoride ion is a very good/poor reducing agent. (Recall the similar relationship between the strengths of conjugate acid and base: if the acid is strong, the base is very $\underline{\hspace{1cm}}$.)	difficult
49. If we want to give numerical expression to the strength of a reducing agent, we could use an equilibrium constant as we did for acids and bases. It is a more common practice, however, to express strengths by standard reduction potentials $E°$. For the halide ions they are F^-, 2.65 V; Cl^-, 1.36 V; Br^-, 1.065 V; I^-, 0.535 V The smaller the value of $E°$, the stronger the reducing agent. The best reducing agent of the four halide ions is therefore $\underline{\hspace{1cm}}$.	poor weak
50. Only iodide ion, among the halides, is a sufficiently good reducing agent to produce Prussian blue in ET-3. To get an equation for the overall reaction we combine the partial equations Reduction: $K^+ + Fe^{3+} + Fe(CN)_6{}^{3-} + e^- \rightarrow$ $\underline{\hspace{3cm}}$ Oxidation: $2I^- \rightarrow \underline{\hspace{1cm}} + 2e^-$	I^-
51. The quantity of oxidation has to match that of reduction, or, in other words, the number of electrons gained in one partial equation must match the number lost in the other. To accomplish this we have to multiply the first partial equation by $\underline{\hspace{1cm}}$, including every formula in it, before adding it to the second.	$KFeFe(CN)_6$ (frame 35) I_2

52. In making this combination we can cancel the electrons, because there are the same number on both sides, and we obtain for the final equation $$2K^+ + 2Fe^{3+} + \underline{\hspace{2cm}} + 2I^- \rightarrow$$ $$2KFeFe(CN)_6 + I_2$$	2
53. If this equation is balanced, the net ionic charges must match on the two sides. On the right-hand side there are no ions indicated, so the net charge is zero. On the left-hand side we have $2+$ for $2K^+$ plus _____ for $2Fe^{3+}$ plus $6-$ for $2Fe(CN)_6{}^{3-}$ plus _____ for $2I^-$, making a net charge of zero, the same as for the right-hand side.	$2Fe(CN)_6{}^{3-}$
	$6+$ $2-$

5.10   Test for Strong Reducing Substances. Oxidation of Sulfide and Sulfite (ET-3)

54. Sulfide and sulfite ions are good reducing agents. To predict what they might be oxidized to, we need to know

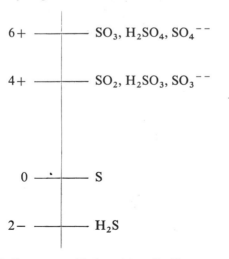

Fig. 5.1. Common oxidation states of sulfur.

the common oxidation states of sulfur (Fig. 5.1). With a mild oxidizing agent like ferricyanide ($E° = 0.356$ V), sulfide ion will be carried up only to the next higher state:

$$S^{--} \rightarrow S + \underline{\hspace{1cm}} e^-$$

55. The test is done in acidic solution. Sulfide ion is a very strong base (Sec. 5.4), so that in acidic solution almost all of it will be in the form of H_2S, a strong/weak acid.	2
56. If we show H_2S as the form in which sulfide occurs, the partial equation is out of balance: $$H_2S \rightarrow S + 2e^-$$ It has two hydrogen atoms on the _____ side that are not matched on the right.	weak
57. What kind of hydrogens shall we put on the right-hand side to effect a balance? Not H atoms; they are not a stable chemical product. Not hydrogen molecules either. True, we could write $H_2S \rightarrow S + H_2$, and this reaction does occur when hydrogen sulfide gas is heated. But note that oxidation of sulfide is then accompanied by reduction of hydrogen from $1+$ to _____. That means we have a complete equation, not a partial equation showing loss of electrons. No electrons would be liberated to cause reduction of ferricyanide.	left-hand
58. The solution to the problem of balancing the hydrogen atoms is evident: H_2S differs from S^{--} by $2H^+$, so we add _____ to the right-hand side of the partial equation $$H_2S \rightarrow S + 2H^+ + 2e^-$$	zero

59. Note that the ionic charges in the partial equation now balance:
   Left side: No ions, ionic charge $=$ _____
   Right side: Zero for S plus $2+$ for $2H^+$ plus _____
   for $2e^-$ gives a net charge of zero.
	$2H^+$

60. To get a complete equation for the action of sulfide ion in ET-3, we combine the partial equation in frame 58 with that for reduction of ferricyanide and formation of Prussian blue in frame 35. The latter must be multiplied by 2 before the combination so that the number of electrons gained in it is equal to the number of electrons _____ in the oxidation of sulfide.	zero $2-$
61. The complete equation is $$2K^+ + \text{_____} + 2Fe(CN)_6^{3-} + H_2S \rightarrow$$ $$2KFeFe(CN)_6 + S + 2H^+$$	lost

62. Making a final check, we find that the net ionic charge on the left is _____ and that on the right is _____, a match.	$2Fe^{3+}$
63. Sulfite ion, according to the plan of the oxidation states of sulfur in Fig. 5.1, will be oxidized to sulfate. The oxidation number of sulfur is therefore increased from $4+$ to _____.	$2+$ $(=2+6-6)$ $2+$
64. We proceed to derive a partial equation for the oxidation of sulfite to sulfate. The test ET-3 is carried out in a solution acidified with HCl. Any sulfite ions will be converted by the strong acid into the strong/weak acid H_2SO_3, which decomposes to water and _____.	$6+$
65. Sulfuric acid, on the other hand, is a strong/weak acid, and very little sulfate will be converted to its conjugate acid.	weak SO_2
66. Our skeleton partial equation, the starting point, is therefore $$SO_2 \rightarrow SO_4^{--}$$ In aqueous acid medium, H_2O and H^+ abound. When it comes to balancing this partial equation, we can count on the availability of H_2O and _____ as common currency for balancing atoms.	strong (or moderately strong if we consider HSO_4^-)
67. Checking the sulfur atoms first, we see that they are already in balance: _____ S on each side.	H^+
68. As a general rule, it is easier to balance oxygen atoms next, then hydrogen atoms. In the skeleton partial equation (frame 66), we have two oxygen atoms on the left and _____ on the right, so there are two extra oxygen atoms on the right for which we have to compensate.	1
69. We can't add two oxygen atoms to the left to get balance, for atomic oxygen is such a reactive/unreactive species that we do not expect to find it in solution at an appreciable concentration. That kind of currency is not in our bank.	four

70. Can we use molecular oxygen to supply the two oxygen atoms? It is certainly available as long as the solution is exposed to air. The trouble is that it is an oxidizing agent itself. The reaction would be $$O_2 + 2SO_2 + 2H_2O \rightarrow 2SO_4^{--} + 4H^+$$ This shows oxidation of SO_2 at the expense of reduction of _____. This is a complete equation, not a _____ equation showing loss of electrons. No electrons would be liberated to cause reduction of ferricyanide (see frame 57).	reactive
71. Nor can we invoke O^{--} ions to balance oxygens. The oxide ion is strongly basic, so in acid solution it will be almost completely converted to water $$O^{--} + \text{_____} \rightarrow H_2O$$ Therefore, oxide ions in aqueous solution are as rare as five-dollar gold pieces, not common currency.	O_2 partial
72. What we do have in solution is lots of water, so we use it: $2H_2O$ supplies _____ oxygen atoms.	$2\,H^+$
73. Adding $2H_2O$ to the left-hand side of the skeleton partial equation, we get $$2H_2O + SO_2 \rightarrow \text{_____}$$ There are now _____ oxygen atoms on both sides.	two
74. To balance hydrogen atoms we again use common currency, namely, H^+ ions: $$2H_2O + SO_2 \rightarrow SO_4^{--} + \text{_____} H^+$$	SO_4^{--} four
75. Having balanced the atoms in the partial equation, we next balance ionic charges. On the left-hand side no ions are shown, so the net charge is _____. On the right-hand side the ionic charges are _____ and _____ for a net charge of $2+$.	4
76. To balance ionic charges we have to add two electrons to the left/right -hand side.	zero $2-$ $4+$

77. The balanced partial equation is $$2H_2O + SO_2 - \underline{\hspace{2cm}} + 4H^+ + 2e^-$$	right
78. This must be combined with the partial equation for reduction of ferricyanide and formation of Prussian blue (frame 35) to get the complete equation for the behavior of sulfite in ET-3: $$2K^+ + 2Fe^{3+} + 2Fe(CN)_6{}^{3-} + 2H_2O + SO_2 \rightarrow$$ $$\underline{\hspace{3cm}} + SO_4{}^{--} + 4H^+$$	$SO_4{}^{--}$
	$2KFeFe(CN)_6$

At this point it may be useful to recapitulate the sequence of steps in balancing partial equations.

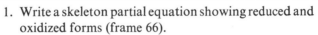

> 1. Write a skeleton partial equation showing reduced and oxidized forms (frame 66).
> 2. Balance atoms.
>   (a) Balance first the atoms that undergo change in oxidation number (frame 67).
>   (b) Balance oxygen atoms next. In acidic solutions add H_2O to the side deficient in oxygen, 1 H_2O for each O required (frames 72 and 73). In basic solutions use 2 OH^- for each O required.
>   (c) Balance hydrogen atoms. In acidic solution use H^+ for this purpose (frame 74). In basic solution add H_2O and release OH^-.
> 3. Balance ionic charges by adding electrons to the side that is too positive or insufficiently negative (frames 75 and 76).

5.11   Test for Strong Reducing Substances. Oxidation of Nitrite (ET-3)

79. The common oxidation states of nitrogen are shown in Fig. 5.2. The position of nitrite ion or its protonated form in acid solution, nitrous acid, indicates that it can be either oxidized or reduced. Here we consider only oxidation; reduction is dealt with in Sec. 5.12.

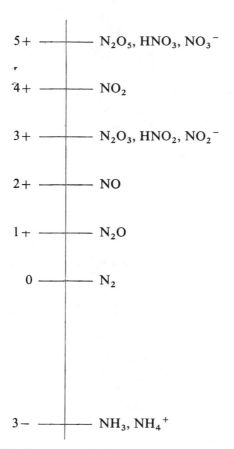

Fig. 5.2. Common oxidation states of nitrogen.

Nitrite might be oxidized to NO_3^- or to _____, for either form contains nitrogen in a higher oxidation state than _____, the value it has in nitrite.

80. Many oxidizing agents, including ferricyanide, carry nitrite all the way to nitrate, so the skeleton partial equation, our starting point, is $$HNO_2 \rightarrow \text{_____}$$	NO_2 $3+$
81. We write the formula HNO_2, rather than NO_2^-, because nitrous acid is a strong/weak acid and will be formed by protonation of the ion in the acid solution of ET-3. We write NO_3^-, not HNO_3, because nitric acid is a strong/weak acid and remains ionic.	NO_3^-

82. We note that nitrogen atoms are/are not in balance, so we begin by balancing _____ atoms.	weak strong
83. The skeleton $HNO_2 \rightarrow NO_3^-$ has an extra oxygen atom on the right, so we add _____ to the left. Hydrogen atoms are then balanced by adding _____ to the right-hand side.	are oxygen
84. The partial equation balanced with respect to atoms is then	H_2O $3\,H^+$
85. Finally, we balance _____ charges by adding electrons to the left/right -hand side: $$H_2O + HNO_2 \rightarrow NO_3^- + 3H^+ + \text{\_\_\_\_\_}$$	$H_2O + HNO_2 \rightarrow NO_3^- + 3H^+$
86. Combining this partial equation with that for reduction of ferricyanide and formation of Prussian blue (frame 35), we get the complete equation for the behavior of nitrite in ET-3.	ionic right $2e^-$
	$2K^+ + 2Fe^{3+}$ $+ 2Fe(CN)_6^{3-}$ $+ H_2O + HNO_2 \rightarrow$ $2KFeFe(CN)_6$ $+ NO_3^- + 3H^+$

5.12   Test for Strong Oxidizing Agents (ET-4)

87. Let us review a bit. An oxidizing agent (1) causes _____ of something else, (2) has an atom in it that undergoes a _____ in oxidation number, and (3) _____ electrons from a reducing agent.	
88. Of the anions we are considering, F^- ion, for one, is already in its lowest oxidation state. Its oxidation number cannot be less than _____, so it cannot function as an oxidizing agent.	oxidation decrease gains

89. The other halide ions, _____, _____, and _____, are in a similar position, and another simple anion, namely, _____, can be reduced no further.

    All the other anions we are considering could, in principle, be oxidizing agents, but under the conditions of ET-4 the only active ones are nitrate and nitrite.

$1-$

---

90. A glance at the oxidation states of nitrogen (Fig. 5.2) reveals the bewildering number of possible reduction products of nitrate ion. Chemical theory is no help in predicting the actual product obtained—we have to resort to laboratory experience. It is not uncommon to find more than one reduction product produced by a given reducing agent.

    Here are some empirical rules as a guide.

> 1. Concentrated (16 M) nitric acid is reduced largely to NO_2.
> 2. More dilute nitric acid (3–6 M) is reduced largely to NO.
> 3. Very dilute nitric acid is reduced by active metals such as zinc and aluminum to ammonium ion, $NH_4{}^+$.
> 4. Nitrate ion in basic solution is reduced by metals such as zinc and aluminum to ammonia, NH_3.

    In ET-4, nitric acid in dilute solution is reduced by manganese(II) in the presence of concentrated hydrochloric acid. We therefore invoke rule (2) and write as a starter

$$NO_3{}^- \rightarrow \text{_____}$$

Cl^-
Br^-
I^-
S^{--} (see Fig. 5.1)

---

91. Now see if you can balance atoms and charges to get a complete partial equation:

    _____

NO

---

92. The oxidation of manganese(II) under the conditions of ET-4 is not completely elucidated. Chloro complexes of manganese in oxidation states II and III are probably involved. For simplicity, we write the partial equation for oxidation as $Mn^{++} \rightarrow Mn^3 + \text{_____}$.

$3e^- + 4H^+ + NO_3{}^- \rightarrow$
$\qquad NO + 2H_2O$

93. Now combine the partial equations for oxidation and reduction to get the complete equation for the behavior of nitrate in ET-4: $4H^+ + NO_3^- + \underline{\hspace{1cm}}Mn^{++} \rightarrow$ $\underline{\hspace{1cm}}Mn^{3+} + NO + 2H_2O$	e^-
94. Nitrous acid, like nitrate, will be reduced to NO under the conditions of ET-4. Derive a partial equation for the reduction:	3, 3
95. Now put together the partial equation for reduction of nitrous acid with that for oxidation of manganese(II) to get the complete equation for the behavior of nitrite in ET-4.	$e^- + HNO_2 + H^+ \rightarrow$ $H_2O + NO$
	$Mn^{++} + HNO_2$ $+ H^+ \rightarrow Mn^{3+}$ $+ NO + H_2O$

REVIEW EXERCISES

5.6. What is the oxidation number of a sulfur atom in the dithionite ion, $S_2O_4^{--}$? Can this substance conceivably be an oxidizing agent? It can function as a reducing agent; what does that mean? What would be a plausible form to which it might be oxidized in basic solution?

5.7. What is the oxidation number of manganese in potassium permanganate, $KMnO_4$? Can this substance function as a reducing agent? It is an oxidizing agent; what does that mean? If its reduction product in acidic solution is Mn^{++} ion, write a balanced partial equation for the reduction half-reaction.

5.8. Criticize the following partial equations, which are intended to represent half-reactions in acid solution:

(a) $S^{4+} \rightarrow S^{6+} + 2e^-$

(b) $H_3PO_3 + O^{--} \rightarrow H_3PO_4 + 2e^-$

(c) $NO_3^- + 7H^+ + 8e^- \rightarrow NH_4^+ + 3OH^-$

5.9. Balance the following partial equations:

(a) $NO_3^- \rightarrow NO_2$ (b) $NO_3^- \rightarrow NH_4^+$ (c) $NO_3^- \rightarrow NH_3$

Under what conditions does each change take place?

5.10. Separate the following into partial equations and balance by the ion–electron method:

$$HNO_2 + HNO_2 \rightarrow NO + NO_3^-$$

5.11. Sulfurous acid and hydrogen sulfide react in acid solution. What would you expect to be the sulfur-containing product of the reaction? Derive a balanced equation from partial equations for the oxidation and reduction.

PRECIPITATION REACTIONS

5.13 Precipitation of Insoluble Silver Salts (ET-5)

1. Salts of nitric acid contain the nitrate ion, which has a net charge of _____ spread over three oxygen atoms.	
2. The ratio of charge to volume (charge density) is low for NO_3^- ion, and its attraction for cations is comparatively strong/weak.	$1-$
3. The solubility rule for nitrates states that all nitrates are insoluble/soluble.	weak
4. This is because the attraction of water dipoles for the ions in crystals of metal nitrates is sufficient to overcome the strong/weak forces bonding the ions in the crystals.	soluble
5. Silver nitrate, on the basis of the general rule, should be insoluble/soluble in water. Moreover, because it is a salt, we expect it to be a strong/weak electrolyte.	weak
6. A 0.2 M solution of silver nitrate will therefore contain high/low concentrations of nitrate and silver ions: specifically, $[Ag^+] = [NO_3^-] = $ _____ M.	soluble strong
7. Halide ions, like nitrate, have a net charge of _____ unit, but they are more compact than nitrate because they are monatomic. We expect them to attract cations less/more strongly than nitrate does. [In addition to electrostatic attraction, there is some covalent bonding between silver and halide ions, especially iodide.]	high 0.2
8. The solubility rule for chlorides, bromides, and iodides states that most of their salts are insoluble/soluble.	$1-$ more
9. Even so, there are more insoluble halides than nitrates. The chief insoluble chlorides, for example, are $PbCl_2$, Hg_2Cl_2, and _____.	soluble

10. We expect a salt like sodium bromide to be insoluble/ soluble in water: its ions tend to remain as hydrated units in solution instead of clustering in a crystal.	AgCl
11. Very few salts are weak electrolytes; we expect sodium bromide to be a strong/weak electrolyte: its ions remain separate in solution instead of bonding covalently into molecules.	soluble
12. Even insoluble salts may be strong electrolytes. Their ions may cluster to give a precipitate, yet fail to bond together covalently to give _____ in solution.	strong
13. We expect a 0.2 M solution of sodium bromide, a strong, soluble electrolyte, to contain ions at high concentration: $[Na^+] = 0.2\ M = [\underline{\hspace{1cm}}]$.	molecules
14. Precipitation reactions occur when two ions are brought together at concentrations high enough to give a super-saturated solution. Such solutions contain higher/lower concentrations of ions than can remain in equilibrium with the solid salt.	Br^-
15. Precipitation restores equilibrium by removing excess ions from solution to give the _____. The solution that remains after precipitation is complete is at equilibrium with the solid salt and must be saturated/ unsaturated.	higher
16. The equilibrium condition for a saturated solution is expressed by the *solubility product*. Equilibrium holds for silver bromide, for example, when $$K_{sp} = [Ag^+][Br^-]\quad\text{(equilibrium)}$$ This expresses the requirement that the product of con-centrations of the ions must equal a fixed value K_{sp}. So if one ion concentration is increased—for example, by adding more silver nitrate or sodium bromide to the saturated solution—the other ion concentration must decrease/increase to maintain the product equal to K_{sp}. In a saturated solution the two ion concentrations do/do not have to be equal.	precipitate (solid salt) saturated

17. On this basis, a supersaturated solution is one for which the instantaneous concentration product, $[Ag^+][Br^-]$, is greater/less than the equilibrium value, K_{sp}.

| decrease |
| do not |

18. Because supersaturation is normally followed by precipitation, the condition for precipitation of silver bromide is

$$[Ag^+][Br^-] > K_{sp} \quad \text{(precipitation)}$$

In words, the product of concentrations of its ions must be larger/smaller than the equilibrium value, K_{sp}.

greater

19. This condition is established immediately when we mix equal volumes of 0.2 M solutions of silver nitrate and sodium bromide, because at the instant of mixing the product of concentrations of the _____ and _____ ions, namely, 0.1×0.1, is very large in comparison with the value of K_{sp} for AgBr, which is 5.2×10^{-13}.

[The silver and bromide ion concentrations are 0.1 M, not 0.2 M, because the solutions dilute each other when they are mixed.]

larger

20. To write an equation for this reaction, we recall that silver nitrate and sodium bromide are both soluble, strong electrolytes, so we use ionic/molecular formulas for these salts.

$$Ag^+ + NO_3^- + Na^+ + Br^- \rightarrow$$
$$Na^+ + NO_3^- + \text{_____}$$

| Ag^+ |
| Br^- |

21. This equation shows two ions unchanged by the reaction, namely, Na^+ and _____. The essential reaction is precipitation of silver bromide:

$$Ag^+ + \text{_____} \rightarrow AgBr(s)$$

| ionic |
| AgBr(s) |

22. A solubility rule states that virtually all sodium salts are insoluble/soluble in water. Thus we expect sodium sulfide to be a soluble salt. It should also be a strong/weak electrolyte.

| NO_3^- |
| Br^- |

23. When 0.2 M solutions of silver nitrate and sodium sulfide are mixed, silver sulfide precipitates. The product of ion concentrations at the instant of mixing must be greater than the equilibrium value:

$$[Ag^+]^2[S^{--}] > \text{_____}$$

| soluble |
| strong |

24. This condition must be fulfilled if precipitation is to occur and indicates that at the instant of mixing the solution is <u>supersaturated/unsaturated</u> in silver sulfide.	K_{sp}
25. An equation for the essential reaction between sodium sulfide and silver nitrate is	supersaturated
26. Precipitation of a number of other insoluble silver salts may occur when high concentrations of other anions react with silver ion in neutral or basic solution: $$3Ag^+ + PO_4{}^{3-} \rightarrow \underline{\hspace{3cm}}$$ $$\underline{\hspace{2cm}} + SO_3{}^{--} \rightarrow Ag_2SO_3(s)$$ $$2Ag^+ + \underline{\hspace{2cm}} \rightarrow Ag_2CO_3(s)$$	$2Ag^+ + S^{--} \rightarrow$ $Ag_2S(s)$ (Na^+ and $NO_3{}^-$ ions are bystanders)
	$Ag_3PO_4(s)$ $2Ag^+$ $CO_3{}^{--}$

5.14   Dissolution of Insoluble Silver Salts (ET-5)

27. The silver salts of chloride, bromide, iodide, and sulfide ions differ from other silver salts in being insoluble in dilute nitric acid. Anions such as $PO_4{}^{3-}$, $SO_3{}^{--}$, and $CO_3{}^{--}$ are <u>acidic/basic</u> and react with hydrogen ions to form <u>strong/weak</u> acids.	
28. The reactions of these anions with nitric acid are represented by $$PO_4{}^{3-} + 3H^+ \rightarrow H_3PO_4$$ $$SO_3{}^{--} + 2H^+ \rightarrow H_2SO_3 \rightarrow H_2O + \underline{\hspace{1.5cm}}$$ $$CO_3{}^{--} + 2H^+ \rightarrow \underline{\hspace{1.5cm}} \rightarrow H_2O + CO_2(g)$$	basic weak
29. These reactions reduce the concentration of each anion so much that the product of concentrations of anion and silver ion falls below the equilibrium value; for example, $$[Ag^+]^3[PO_4{}^{3-}] < \underline{\hspace{1.5cm}}$$	$SO_2(g)$ H_2CO_3

30. Such a solution is saturated/unsaturated, and some of the silver phosphate dissolves in an attempt to restore equilibrium.	K_{sp}
31. Each example of acid–base reaction followed by dissolution of the precipitate can be summarized in a single equation: $Ag_3PO_4(s) + 3H^+ \rightarrow$ _____ $+ H_3PO_4$ $Ag_2SO_3(s) + 2H^+ \rightarrow 2Ag^+ + H_2O +$ _____ $Ag_2CO_3(s) + 2H^+ \rightarrow$ _____	unsaturated
32. Chloride, bromide, and iodide ions are anions of strong/weak acids. Their basic strength in dilute solution is virtually _____.	$3Ag^+$ $SO_2(g)$ $2Ag^+ + H_2O + CO_2(g)$
33. Since these anions do not react with protons, their concentrations are unaffected by nitric acid. The halides AgCl, AgBr, and AgI do/do not dissolve in nitric acid.	strong nil (zero)
34. Sulfide ion, on the other hand, is a strongly basic anion (Sec. 5.4). Why, then, is silver sulfide insoluble in dilute nitric acid? The answer lies in its extremely low solubility product, $K_{sp} = 7 \times 10^{-50}$. In a competition between Ag^+ and H^+ for S^{--}, the silver ion nearly always loses/wins.	do not
35. The equilibrium constant for $Ag_2S(s) + 2H^+ \leftrightharpoons H_2S + 2Ag^+$ has the very small value 7×10^{-28}. That means that when all solutes are at approximately unit concentration this reaction does/does not occur to an appreciable extent.	wins
36. Silver chloride and silver bromide can be dissolved by aqueous ammonia. Silver ion and ammonia molecules form a weakly dissociated complex $Ag^+ + 2NH_3 \leftrightharpoons Ag(NH_3)_2^+ \qquad \beta_2 = 1.6 \times 10^7$ The large value of the equilibrium constant β_2 indicates that when excess ammonia is used, virtually all/none of the silver ions will be removed from the field of action.	does not

37. Because of the decrease in silver ion concentration, the condition is established $$K_{sp} > [Ag^+][Cl^-]$$ which corresponds to a solution that is saturated/unsaturated.	all
38. The precipitate dissolves in an effort to restore equilibrium: $$AgCl(s) + 2NH_3 \rightarrow \underline{\hspace{3cm}} + Cl^-$$	unsaturated
39. Silver bromide has a lower solubility product than silver chloride and requires a higher concentration of ammonia to bring about the dissolution. Silver iodide is not dissolved by ammonia, for it is the least/most soluble of the silver halides.	$Ag(NH_3)_2{}^+$
	least

5.15 Precipitation and Dissolution of Insoluble Calcium Salts (ET-6)

40. The reagent calcium nitrate is insoluble/soluble in water and is a strong/weak electrolyte. In solution it provides high concentrations of calcium and _____ions. Only the calcium ions are of interest in the following reactions.	
41. Because of its double charge, calcium ion forms strong/weak electrostatic bonds with the very small univalent fluoride ion and with multiply charged ions like carbonate.	soluble strong nitrate
42. Typical reactions of calcium ion include $$Ca^{++} + 2F^- \rightarrow \underline{\hspace{2cm}}$$ $$Ca^{++} + CO_3{}^{--} \rightarrow CaCO_3(s)$$ $$Ca^{++} + \underline{\hspace{3cm}} \rightarrow CaSO_3(s)$$ $$3Ca^{++} + 2PO_4{}^{3-} \rightarrow \underline{\hspace{2cm}}$$	strong

43. All four of these anions are basic, so we expect the precipitates to be insoluble/soluble in strong acids. Since they differ in basic strength, we can separate them by using a weak acid, which gives a high/low concentration of hydrogen ion.	$CaF_2(s)$ SO_3^{--} $Ca_3(PO_4)_2(s)$
44. Acetic acid is a strong/weak acid and will dissolve all the calcium salts in frame 42 except calcium fluoride, which contains the least basic anion.	soluble low
45. In writing equations for the dissolution, we use an ionic/molecular formula to represent the weak acid: $CaCO_3(s) + 2HOAc \rightarrow$ $\quad\quad Ca^{++} + 2OAc^- + H_2O + \underline{\quad\quad}$ $CaSO_3(s) + 2HOAc \rightarrow$ $\quad\quad \underline{\quad\quad} + \underline{\quad\quad} + H_2O + SO_2(g)$	weak
46. These reactions are driven to completion by formation of the gaseous \underline{\quad\quad}.	molecular $CO_2(g)$ $Ca^{++}; 2OAc^-$
47. Because HOAc is a weaker acid than H_3PO_4, it does not completely protonate PO_4^{3-} ions: $Ca_3(PO_4)_2(s) + 4HOAc \rightarrow$ $\quad\quad 3Ca^{++} + 2H_2PO_4^- + \underline{\quad\quad}$	oxides
48. When this solution of calcium phosphate in acetic acid is subsequently made basic with ammonia, $H_2PO_4^-$ ion behaves as a proton acceptor/donor toward NH_3: $H_2PO_4^- + 2NH_3 \rightarrow PO_4^{3-} + \underline{\quad\quad}$	$4OAc^-$
49. The resulting increase in phosphate ion concentration establishes the condition for precipitation: $[Ca^{++}]^3[PO_4^{3-}]^2 \underline{\quad\quad} K_{sp}$	donor $2NH_4^+$
50. The net reaction is represented by $3Ca^{++} + 2H_2PO_4^- + 4NH_3 \rightarrow Ca_3(PO_4)_2(s) + \underline{\quad\quad}$	$>$
	$4NH_4^+$

5.16 Precipitation of Insoluble Barium Salts (ET-7)

51. Barium nitrate is insoluble/soluble in water and is a strong/weak electrolyte. It therefore supplies a high/low concentration of barium ion.	
52. Sulfate ion is a very weak base, so in acidic solution only a small proportion of sulfate will be converted to its conjugate acid _____.	soluble strong high
53. An acidic solution of a sulfate therefore usually contains a sufficiently high concentration of sulfate ion to establish the condition for precipitation of barium sulfate: $$[Ba^{++}][SO_4^{--}] \text{_____} K_{sp}$$	HSO_4^-
54. Other anions that might precipitate with barium ion are more basic than sulfate and are sequestered or destroyed by acid: $$BaF_2(s) + 2H^+ \rightarrow Ba^{++} + \text{_____}$$ $$BaSO_3(s) + 2H^+ \rightarrow Ba^{++} + H_2O + \text{_____}$$ $$BaSO_4(s) + H^+ \rightarrow \text{virtually no reaction}$$	$>$
55. Note that sulfites are often contaminated by sulfate formed by slow air oxidation: $$O_2 + 2SO_3^{--} \rightarrow 2SO_4^{--}$$ For that reason, barium nitrate will often produce a precipitate when added to a sulfite solution, and the precipitate is not soluble in acid. It is not barium sulfite but barium _____.	2HF $SO_2(g)$
	sulfate

REVIEW EXERCISES

5.12. Write equations for the following reactions:
(a) Silver nitrate with nickel(II) iodide
(b) Barium fluoride with hydrochloric acid
(c) Silver phosphate with aqueous ammonia

5.13. Criticize the following equations:

(a) $2AgNO_3 + MgCl_2 \rightarrow 2AgCl + Mg(NO_3)_2$

(b) $Ca^{++} + (NO_3)_2^{--} + Na_2^{++} + CO_3^{--} \rightarrow CaCO_3(s) + 2NaNO_3$

(c) $BaCO_3(s) + H^+ \rightarrow Ba^{++} + HCO_3^-$

5.14. Three possible conditions may be applicable to solutions of silver carbonate

$$[Ag^+]^2[CO_3^{--}] > K_{sp}; \quad K_{sp} = [Ag^+]^2[CO_3^{--}]; \quad [Ag^+]^2[CO_3^{--}] < K_{sp}$$

Which condition corresponds to a saturated solution? an unsaturated solution? a supersaturated solution? Which is the condition for forming a precipitate of silver carbonate? Which for dissolving it?

5.15. Concentrated sulfuric acid is largely molecular H_2SO_4, which is a much better proton donor than the hydrated hydrogen ion present in dilute acids. Account for the fact that barium sulfate will dissolve in the concentrated acid. Write an equation for the reaction.

Detection

of the Cation

6.1 Introduction

Simple substances contain only one positive ion. Most are salts containing ammonium ion or the cation of a metal. Detection of the following cations[1] is discussed in this chapter:

Al^{3+}, $NH_4{}^+$, Sb(III), As(III), Ba^{++}, Bi^{3+}, Cd^{++}, Ca^{++}, Cr^{3+}, Co^{++}, Cu^{++}, Fe^{++}, Fe^{3+}, Pb^{++}, Mg^{++}, Mn^{++}, $Hg_2{}^{++}$, Hg^{++}, Ni^{++}, K^+, Ag^+, Na^+, Sr^{++}, Sn^{++}, Sn(IV), and Zn^{++}

We consider first the initial examination of the solid sample. Then a solution of it is prepared, and various preliminary "crash" tests are made on the solution. From these observations it is usually possible to infer the probable presence of one cation. This hypothesis is then checked by performing a specific identification test. All steps in the analysis from initial examination to identification should be recorded in your laboratory notebook.

---

[1] When the symbol of the element is followed by a roman numeral designating the oxidation state, the element exists as a complex rather than as a simple cation.

6.2   Initial Examination

(a) Physical Appearance of the Sample

Note first whether the substance is coarsely crystalline, finely powdered, or possibly amorphous. Is it colored? A tentative identification may be possible on the basis of the information in Table 6.1. Any such general summary, it must be recognized, is imprecise in its description of colors. Not all the red substances listed in the table are the same shade of red. The deep purplish red of silver chromate is easily distinguished from the orange red of Pb_3O_4 or the brownish red of iron(III) oxide. Moreover, the anhydrous and hydrated forms of a salt may differ profoundly in color; for example, $CuSO_4$ is colorless and $CuSO_4 \cdot 5H_2O$ is blue, $NiSO_4$ is yellow and $NiSO_4 \cdot 6H_2O$ is green. The table of inorganic substances in a chemical handbook may be consulted for individual differences such as these. Any identification based on color must be tentative and needs to be supported by other evidence.

Table 6.1

Common Colored Solid Substances

Color	Substances
Red	Hydrated Mn(II) salts (pink), hydrated Co(II) salts, HgI_2, HgO, HgS (cinnabar), Pb_3O_4, Ag_2CrO_4, AsI_3, Cu_2O, CrO_3, Fe_2O_3 (red brown)
Orange	Sb_2S_3, SnI_2, many dichromates
Yellow	As_2S_3, CdS, HgO, PbI_2, PbO, SnS_2, anhydrous Ni(II) salts, many chromates
Green	Hydrated Fe(II) and Ni(II) salts, Cr_2O_3 (almost black), $CrCl_3 \cdot 6H_2O$ (one form) and other Cr(III) salts, $CuCl_2 \cdot 2H_2O$ (blue green) and certain other Cu(II) salts, anhydrous $CoBr_2$, K_2MnO_4
Purple	$KMnO_4$ (almost black)
Violet	Many Cr(III) salts (very dark) and certain Fe(III) salts such as $Fe(NO_3)_3 \cdot 9H_2O$ (pale)
Blue	Hydrated Cu(II) salts (sometimes yellowish or greenish), anhydrous $CoCl_2$
Brown	$FeCl_3 \cdot 6H_2O$ (yellowish), anhydrous $CuCl_2$, PbO_2, CdO, Bi_2S_3 (dark), Bi_2O_3, SnS
Black	Fe_3O_4, CuO, NiO, MnO_2, CuS, Cu_2S, HgS, PbS, Sb_2S_3 (stibnite), FeS, CoS, NiS, Ag_2S, $CuBr_2$, BiI_3

(b) Flame Tests

Flame tests are likewise helpful, but not necessarily conclusive. Some common examples are given in Table 6.2. To carry out the test you must start with a clean platinum or Nichrome wire. Dip the wire into some 12 M HCl and bring it into

Table 6.2

Common Flame Colors*

Element	Color
Na	Strong, persistent yellow
K	Weak violet, red through cobalt blue glass
Ca	Strong orange red
Sr	Carmine
Ba	Yellow green
Cu	Green
As	Bluish white
Sb	Bluish white
Pb	Pale blue

* The colors given by the salts moistened with 12 M HCl.

the oxidizing part of the flame (above the inner cone). Repeat until no color is imparted to the flame. Traces of solids may be difficult to remove from Nichrome, so it may be necessary to snip off the end of this kind of wire with scissors. Moisten a little of the solid sample with 12 M HCl, dip the clean wire in the mixture, and bring it into the oxidizing part of the flame to obtain the flame color. Flame tests should always be corroborated by other tests.

6.3 Preparation of the Solution

(a) Choice of Solvent

Test the solubility of the sample in water (T-2).[2] If it does not dissolve in water, try 6 M HCl and then 6 M HNO$_3$. The acids may dissolve the solid by suppressing hydrolysis; for example, the reaction

$$SbCl_3 + H_2O \leftrightarrows SbOCl(s) + 2H^+ + 2Cl^-$$

is reversed by addition of hydrochloric acid. Or the acids may remove an anion to form a weak acid, a volatile product, or an oxidized form:

$$Ca_3(PO_4)_2(s) + 4H^+ \rightarrow 3Ca^{++} + 2H_2PO_4^-$$

$$Cu_2(OH)_2CO_3(s) + 4H^+ \rightarrow 2Cu^{++} + 3H_2O + CO_2(g)$$

$$3PbS(s) + 8H^+ + 2NO_3^- \rightarrow 2NO(g) + 3S(s) + 3Pb^{++} + 4H_2O$$

[2] T-2 refers to technique T-2 in Sec. 3.3.

If water or one of the dilute acids does not dissolve your sample, it may be necessary to use a concentrated acid or aqua regia (3 volumes 12 M HCl to 1 volume 16 M HNO_3).

Having chosen a solvent, prepare a solution for analysis by dissolving about 100 mg of sample [T-3(b)] in 5 ml of solvent.

(b) Color Differences Between Solid and Solution

A comparison of colors of the original sample and its solution if often useful. Yellow $NiSO_4$, for example, will dissolve in water to give a green solution containing the same $Ni(H_2O)_6{}^{++}$ ion found in the solid hydrated salts. If hydrochloric acid is the solvent, colored chloro complexes may be formed. The yellow color of $FeCl^{++}$ and $FeCl_2{}^+$ in HCl is particularly distinctive and intense. Finally, there are some cations, which though colorless in solution, form colored solids with anions such as I^-, O^{--}, or S^{--}; for example, yellow PbI_2, AgI, PbO, As_2S_3, CdS; red HgI_2, HgS, Pb_3O_4; black HgS, PbS, BiI_3, $CuBr_2$. The colors are a result of partial covalent bonding between anion and cation in the solid and disappear when the bonding is destroyed by dissolving the solid (Sec. 7.1). Some anions, of course, are themselves colored, yellow chromate ion being a striking example.

(c) Cation–Anion Incompatibilities

If the identity of the anion is established prior to the cation analysis, it is frequently possible to eliminate certain cations from further consideration. If the anion, for example, is sulfate and if the sample is soluble in water, the cation cannot be Pb^{++}, Ba^{++}, Sr^{++}, or Ca^{++}, for these form insoluble sulfates. Nor is it likely to be Bi^{3+}, Sb(III), or Hg^{++}, the sulfates of which hydrolyze and precipitate basic salts, as in the reaction

$$3Hg^{++} + SO_4{}^{--} + 4H_2O \rightarrow Hg_3(OH)_4SO_4(s) + 4H^+$$

(d) Removal of Interfering Anions

The presence of reducing anions may interfere with the tests for some cations. Sulfite, nitrite, or sulfide can be removed by acidifying the solution with 6 M HCl and boiling it (T-6):

$$SO_3{}^{--} + 2H^+ \rightarrow H_2O + SO_2(g)$$

$$3NO_2{}^- + 2H^+ \rightarrow 2NO(g) + NO_3{}^- + H_2O$$

$$S^{--} + 2H^+ \rightarrow H_2S(g)$$

Hydrogen iodide is not so volatile or unstable, so iodide must be removed by oxidation. To the solution of the sample in a casserole, add several drops of 16 M HNO_3 and evaporate the solution almost to dryness (T-7). A brown color in the solution and violet fumes indicate the evolution of iodine:

$$8H^+ + 2NO_3{}^- + 6I^- \rightarrow 3I_2(g) + 4H_2O + 2NO(g)$$

6.4 Preliminary Tests

It is useful to test the solution of the sample with various reagents that react with sizable numbers of ions. The formation of a precipitate or the absence of apparent reaction indicates the presence or absence of certain groups of ions. The simplicity of the tests makes them worth trying, but the results must be confirmed by other tests. For each test, take one drop of the solution of the sample, dilute with a few drops of water to get a convenient volume, and then add the particular reagent, a drop at a time, mixing thoroughly after each addition (T-5), until no further change is observed.

(a) Action of Excess Sodium Hydroxide

Add 6 M NaOH until an excess is present; the final mixture should be strongly basic—check with litmus (T-12). Centrifuge to consolidate any precipitate; the white, gelatinous hydroxides of magnesium and aluminum are difficult to see when dispersed. If no precipitate remains, the following cations are absent:

$$Ag^+, Hg_2^{++}, Hg^{++}, Bi^{3+}, Cu^{++}, Cd^{++}, Fe^{++}, Fe^{3+}, Mn^{++}, Co^{++}, Ni^{++},$$
$$\text{and } Mg^{++}$$

If there is a precipitate, its color may be significant; for example, blue $Co(OH)_2$ or $Cu(OH)_2$, red brown $Fe(OH)_3$, yellow HgO, and brown Ag_2O. Consult a handbook for the colors of other oxides and hydroxides.

Other observations may or may not be made, depending on circumstances. Oxides or hydroxides of the following will precipitate at first, then redissolve in excess base:

$$Al^{3+}, Sb(III), As(III), Cr^{3+}, Pb^{++}, Sn^{++}, Sn(IV), \text{ and } Zn^{++}$$

Whether or not you observe this depends on the rate of addition of base. It is easy to overshoot the precipitation step. The chromium hydroxo complex is green:

$$Cr_2O_3 \cdot 3H_2O(s) + 2OH^- \rightarrow 2Cr(OH)_4^-$$

Copper(II) and cobalt(II) hydroxides may dissolve if a large excess of base is used. Pink manganese(II) hydroxide, though remaining undissolved, will darken in time, owing to oxidation to brown manganese(III) oxyhydroxide:

$$O_2 + 4Mn(OH)_2(s) \rightarrow 4MnO(OH)(s) + 2H_2O$$

(b) Action of Excess Aqueous Ammonia

Use 15 M NH$_3$ and again check with litmus (T-12) to make sure the final solution is strongly alkaline. If no precipitate forms, all cations are absent except:

Group A: Ag^+, Cu^{++} (deep blue solution), Cd^{++}, Co^{++} (yellow brown solution), Ni^{++} (blue solution), Zn^{++}

Group B: Na^+, K^+, and NH_4^+

The cations in Group A form weakly dissociated ammonia complexes:

$$Ag^+ + 2NH_3 \rightarrow Ag(NH_3)_2{}^+$$

$$Co^{++} + 6NH_3 \rightarrow Co(NH_3)_6{}^{++}$$

$$Zn^{++} + 4NH_3 \rightarrow Zn(NH_3)_4{}^{++}$$

Those in Group B do not form insoluble hydroxides. The hydroxides of Ba^{++}, Sr^{++}, and Ca^{++} are also moderately soluble and should not precipitate, but ammonia solution has often absorbed sufficient carbon dioxide from air to give small precipitates of the carbonates of these ions. The white, gelatinous hydroxides of magnesium and aluminum are difficult to see when dispersed. Always centrifuge before deciding on the presence or absence of a precipitate.

(c) Action of Sulfuric Acid

Use one drop of 6 M H_2SO_4 and add several drops of 95% ethyl alcohol to decrease the solubility of the sulfates, particularly that of calcium. Stir vigorously for a minute to overcome supersaturation. If a precipitate forms, the presence of one of the following is indicated:

$$Pb^{++}, Ba^{++}, Sr^{++}, Ca^{++}$$

(d) Action of Thioacetamide in Acidic Solution

Add 6 M NH_3 to the sample until it is basic to litmus or to wide-range indicator paper (T-12). Then add 2 M HCl drop by drop until the pH is between 2 and 3 according to the wide-range paper. If the solution was made too acidic, bring it back with a fraction of a drop of 6 M NH_3 [T-4(b)]. Then add exactly 0.35 ml of 2 M HCl, using a calibrated dropper. Add a few drops of thioacetamide solution and enough water to make a total volume of 2.5 ml. Mix well [T-5(b,c)]. Ignore the formation of a precipitate, if any, for the moment. Check the acidity of the mixture against short-range indicator paper; it should have a pH of 0.5 corresponding to a hydrogen ion concentration of 0.3 M (Secs. 2.4 and 2.5).

Warm the solution in a bath of vigorously boiling water [Fig. 3.3(a)] for 5 minutes. If no precipitate forms, proceed to the next test (e). Thioacetamide hydrolyzes to give hydrogen sulfide (Sec. 2.5):

$$CH_3C(S)NH_2 + H_2O \rightarrow CH_3C(O)NH_2 + H_2S$$

which precipitates sulfides of the more insoluble class in acidic solution (Secs. 2.4–2.6); for example,

$$2Bi^{3+} + 3H_2S \rightarrow Bi_2S_3(s) + 6H^+$$

The possible products in 0.3 M HCl are yellow CdS, As_2S_3, or SnS_2; orange Sb_2S_3 (sometimes CdS); red HgS; brown Bi_2S_3 or SnS; and black HgS, PbS, or CuS. If the acidity was not adjusted properly, CdS may not precipitate (it will appear in the next step), or the sulfides normally precipitated in basic solution (see next procedure) may form.

If the precipitate is yellow or orange, centrifuge (T-8), and withdraw and discard the solution (T-9). Add 10 drops of 0.5 M KOH and heat briefly in the water bath. If the precipitate dissolves, it may be one of the acidic sulfides:

$$SnS_2(s) + OH^- \rightarrow SnS_2OH^-$$

$$As_2S_3(s) + 6OH^- \rightarrow AsS_3{}^{3-} + AsO_3{}^{3-} + 3H_2O$$

$$Sb_2S_3(s) + 6OH^- \rightarrow SbS_3{}^{3-} + SbO_3{}^{3-} + 3H_2O$$

Yellow cadmium sulfide is basic and will not dissolve.

To the alkaline solution of the sulfides add one drop of thioacetamide solution and then carefully add 2 M HCl until the solution is just acid to litmus (T-12). An orange precipitate is Sb_2S_3. A yellow precipitate may be either As_2S_3 or SnS_2. Centrifuge and then withdraw the supernatant liquid. Add to the precipitate 5 drops of 12 M HCl, stir and warm briefly; SnS_2 will dissolve, As_2S_3 will not:

$$SnS_2(s) + 4H^+ + 6Cl^- \rightarrow SnCl_6{}^{--} + 2H_2S(g)$$

(e) Action of Thioacetamide in Basic Solution

If no precipitate of a sulfide was obtained in acidic solution, make the solution alkaline with 6 M NH$_3$ (check with litmus). A white precipitate is ZnS or Al(OH)$_3$; a pink precipitate is MnS; yellow CdS may come down if the acidity was not adjusted properly in (d); a green precipitate is Cr(OH)$_3$; FeS, CoS, and NiS are black. The sulfides are more soluble than those precipitated from acid solution (Sec. 2.4), and the two hydroxides precipitate because the solution is basic (Sec. 2.7).

6.5   Identification Tests

Time devoted to pondering the results of the initial examination, solubility tests, and preliminary tests is better spent than that expended on aimless trials of the following identification tests. In some cases, further identification may be unnecessary or at most equivalent to holding up jeans with both belt and suspenders.

The ions are listed alphabetically by name; for example, antimony (Sb) before arsenic (As). For each ion, preliminary indications of its presence are summarized, an identification test is described, and some possible problems or confusions that might lead to false identification are mentioned. If an identification test gives dubious results, try it on an authentic solution of the ion for comparison. Look for decisive tests, not weak colors or tiny precipitates.

Aluminum(III) Ion

(a) *Preliminary Indications.* Aluminum salts are colorless unless the anion is colored. White, gelatinous Al(OH)$_3$ is precipitated by bases and redissolves in 6 M NaOH (but not in NH$_3$) to give the aluminate ion, Al(OH)$_4{}^-$.

(b) *Identification Test.* To a few drops of a solution of the salt, add 2 drops of aluminon reagent (a dye) and 5 drops of NH_4OAc solution. Check the pH with wide-range paper; it should be between 5 and 7. If it is too low, add more ammonium acetate. A red color indicates the presence of aluminum. Then make the mixture barely alkaline with 6 M NH_3 and centrifuge to settle the precipitate of aluminum hydroxide, which has adsorbed the dye (called a "lake"):

$$Al^{3+} + 3NH_3 + 3H_2O \rightarrow Al(OH)_3(s) + 3NH_4^+$$

(c) *Possible Sources of Confusion.* Other metal ions, such as Fe^{3+}, Cr^{3+}, and Pb^{++}, form insoluble hydroxides in ammoniacal solution and may also give a lake. Watch for indications of these ions in the preliminary tests.

Ammonium Ion

(a) *Preliminary Indications.* Ammonium salts are colorless and soluble in water. If the ammonium ion was present, you would have noted the evolution of pungent ammonia when 6 M NaOH was added to the sample:

$$NH_4^+ + OH^- \rightarrow NH_3(g) + H_2O.$$

(b) *Identification Test.* Put a small amount of the solid sample in a 5-ml beaker. Cut a piece of red litmus paper smaller than the diameter of the beaker, moisten it, and attach it to the convex side of the watch glass. Add several drops of 6 M NaOH to the beaker, being careful not to get any on the upper walls, and immediately cover with the watch glass. If litmus paper turns blue within 2 minutes, report the presence of ammonium ion. The reaction in the beaker is the same as that in (a); on the moist litmus the reverse reaction occurs.

(c) *Possible Sources of Confusion.* If NaOH is dropped carelessly on the upper walls of the beaker, it may flow onto the watch glass and litmus and turn the litmus blue. Laboratory air may contain enough ammonia to turn litmus blue slowly—hence the time limit.

Antimony(III) Ion

(a) *Preliminary Indications.* Bluish white flame test. Antimony(III) oxide or oxychloride (SbOCl) is white and soluble in excess NaOH. Orange Sb_2S_3 is soluble in 0.5 M KOH and dilute HCl.

(b) *Identification Test.* No further test should be required.

(c) *Source of Confusion.* Cadmium sulfide is sometimes more orange than yellow, but it is not soluble in 0.5 M KOH.

Arsenic(III) Ion

(a) *Preliminary Indications.* Bluish white flame test. Yellow As_2S_3, soluble in 0.5 M KOH and 6 M NH_3 but not in dilute HCl.

(b) *Identification Test.* No further test is usually required but, if desired, As_2S_3 may be dissolved in NH_3 and H_2O_2, forming arsenate ion. This is precipitated first as white, crystalline $MgNH_4AsO_4$ and then as reddish brown Ag_3AsO_4. Directions are given in Outline 4, Chapter 8.

(c) *Sources of Confusion.* Yellow CdS is not soluble in 0.5 M KOH; yellow SnS$_2$ is soluble in dilute HCl.

Barium(II) Ion

(a) *Preliminary Indications.* Green flame test. White precipitate of BaSO$_4$ with H$_2$SO$_4$.

(b) *Identification Test.* No further tests are necessary, but see (c).

(c) *Sources of Confusion.* The other insoluble sulfates are readily confused with BaSO$_4$. Lead sulfate, however, unlike BaSO$_4$, is soluble in NH$_4$OAc or NaOH. Barium's flame test differentiates it from strontium or calcium. See also Sec. 6.6 for a microscopic test that differentiates barium and calcium sulfates.

Bismuth(III) Ion

(a) *Preliminary Indications.* White, gelatinous Bi(OH)$_3$ insoluble in excess NaOH or NH$_3$. Very dark brown Bi$_2$S$_3$ obtained from acid solution.

(b) *Identification Test.* Prepare some sodium stannite reagent by adding 6 M NaOH drop by drop to one drop of SnCl$_2$ reagent. Mix carefully after each drop, and continue the addition until Sn(OH)$_2$ redissolves:

$$Sn^{++} + 2OH^- \rightarrow Sn(OH)_2(s)$$

$$Sn(OH)_2(s) + OH^- \rightarrow Sn(OH)_3^-$$

At least 2 drops of NaOH should be required. Then add a few drops of this freshly prepared stannite solution to a precipitate of Bi(OH)$_3$. An immediate change to jet black Bi indicates the presence of Bi(III):

$$3OH^- + 3Sn(OH)_3^- + 2Bi(OH)_3(s) \rightarrow 2Bi(s) + 3Sn(OH)_6^{--}$$

(c) *Source of Confusion.* Prove that mercury(II) is absent first, for it too is reduced to a black metal.

Cadmium(II) Ion

(a) *Preliminary Indications.* White Cd(OH)$_2$, insoluble in NaOH, soluble in NH$_3$. Yellow to orange CdS, insoluble in 0.5 M KOH, soluble in dilute HCl.

(b) *Identification Test.* No further test should be required but see (c). A microscopic test is given in Sec. 6.6.

(c) *Sources of Confusion.* Cadmium sulfide is the most soluble of those precipitated in acid solution, and if the acidity is not adjusted properly, it may not appear until the solution is made basic. Yellow As$_2$S$_3$ and SnS$_2$ and orange Sb$_2$S$_3$ are soluble in 0.5 M KOH.

Calcium(II) Ion

(a) *Preliminary Indications.* Orange red flame test. Moderately insoluble CaSO$_4 \cdot 2H_2O$.

(b) *Identification Test.* No further test is usually necessary. A microscopic test for differentiating the sulfate from those of barium and strontium is given in Sec. 6.6.

(c) *Sources of Confusion.* Lead sulfate is also white but dissolves in NH_4OAc or NaOH. Barium and strontium sulfates give granular crystals distinct from the needle-like growths of $CaSO_4 \cdot 2H_2O$.

Chromium(III) Ion

(a) *Preliminary Indications.* Green to deep violet (almost black) crystals and green to violet solutions. Green to grey green hydrated Cr_2O_3, insoluble in NH_3, soluble in NaOH to give the green chromite ion, $Cr(OH)_4^-$.

(b) *Identification Test.* To a few drops of the solution under test in a casserole, add several drops of 16 M HNO_3 and heat to a boil (T-6). Add a few crystals of $KClO_3$ and again bring to a boil. Repeat addition of HNO_3 and $KClO_3$ once. An orange color due to dichromate ion indicates presence of chromium:

$$H_2O + 6ClO_3^- + 2Cr^{3+} \rightarrow Cr_2O_7^{--} + 6ClO_2(g) + 2H^+$$

Make the solution just basic with 6 M NaOH (T-12); the orange color should change to bright yellow due to chromate:

$$Cr_2O_7^{--} + 2OH^- \rightarrow 2CrO_4^{--} + H_2O$$

(c) *Sources of Confusion.* The solid salts are so deeply colored that it is easy to mistake them for a black compound such as a sulfide; the violet color of $Cr(H_2O)_6^{3+}$ is unmistakable in solution. Green chromium(II) salts usually contain ions such as $CrCl_2(H_2O)_4^+$.

Cobalt(II) Ion

(a) *Preliminary Indications.* Light red to dark reddish orange hydrated salts, red or sometimes blue solutions. Blue or pink hydroxide, somewhat soluble in excess NaOH, soluble in NH_3 to give a yellow brown solution. The hydroxide and the ammonical solution darken on standing, due to oxidation to Co(III) by oxygen of air. Black CoS forms in basic or very weakly acidic solutions.

(b) *Identification Test.* To a drop of the solution of the salt, add a few drops of water and several crystals of NH_4SCN. Then add several drops of NH_4SCN dissolved in alcohol–ether mixture and shake. A blue upper layer containing $Co(SCN)_4^{--}$ indicates the presence of cobalt. Alternatively, the microscopic test in Sec. 6.6 may be tried.

(c) *Sources of Confusion.* Manganese(II) salts are also pink but very pale; $Mn(OH)_2$ is insoluble in NaOH and NH_3, and MnS is not black. In the extraction test, traces of Fe^{3+}, a common impurity, may give a deep red color with thiocyanate that could obscure the blue of the cobalt complex. A little solid NaF can be added to the solution to sequester the iron as the colorless, weakly dissociated FeF_6^{3-} complex.

Copper(II) Ion

(a) *Preliminary Indications.* Solutions and hydrated salts are blue due to $Cu(H_2O)_4^{++}$. Green flame test. Blue hydroxide, somewhat soluble in NaOH, soluble in NH_3 to give a deep blue solution of $Cu(NH_3)_4^{++}$. Black sulfide can be precipitated from acid solution.

(*b*) *Identification Test.* The preliminary indications usually suffice. A microscopic test is given in Sec. 6.6.

(*c*) *Sources of Confusion.* The anhydrous salts range from white $CuSO_4$ to black $CuBr_2$ and CuS; their solutions are blue. Nickel forms a blue complex $Ni(NH_3)_6^{++}$, not as intense a color as that of the copper complex; moreover, nickel hydroxide is green.

Iron(II) and Iron(III) Ions

(*a*) *Preliminary Indications.* Solutions of Fe^{++} and its hydrated salts are usually green. Solid iron(III) salts range from yellow $FeCl_3 \cdot 6H_2O$ to violet $Fe(NO_3)_3 \cdot 9H_2O$. In solution, iron(III) forms yellow or brown complexes:

$$Fe^{3+} + Cl^- \leftrightharpoons FeCl^{++}$$

$$Fe^{3+} + H_2O \leftrightharpoons FeOH^{++} + H^+$$

Red brown $Fe(OH)_3$ or hydrated oxide, insoluble in $NaOH$ or NH_3. Iron(II) hydroxide ranges from white to black, depending on the degree of contamination by iron(III). Both ions give FeS on treatment with thioacetamide.

(*b*) *Identification Test.* (1) Test for Fe^{++}. Dissolve one small crystal of $K_3Fe(CN)_6$ in about 20 drops of water. Add a drop of this solution to one of the solution under test. A blue color or precipitate of $KFeFe(CN)_6$ indicates the presence of iron(II). If the color is green, the reagent may be too concentrated; dilute it and try the test again.

(2) Test for Fe^{3+}. Dissolve a few crystals of NH_4SCN in a little water and add a few drops of this reagent to the solution under test. A dense red color, not precipitate, due to $FeSCN^{++}$ indicates the presence of iron(III).

(*c*) *Sources of Confusion.* In the test for iron(II), other ions will give colored precipitates with ferricyanide, but only Fe^{++} gives a blue one. In the test for Fe^{3+}, mere traces of iron(III), a common impurity in reagents, give a discernible pink color. Also, all aged solutions of iron(II) salts give a test for iron(III) because the iron is slowly oxidized by the oxygen of air:

$$4H^+ + O_2 + 4Fe^{++} \rightarrow 2H_2O + 4Fe^{3+}$$

Do not report traces; you are looking for very definite tests. If in doubt as to whether the color is deep enough, try the test on a few drops of a solution prepared by diluting one drop of standard Fe^{3+} solution with 5 ml of water (mix well).

Lead(II) Ion

(*a*) *Preliminary Indications.* Pale blue flame test. White hydroxide, soluble in excess $NaOH$, not in NH_3. White insoluble $PbSO_4$. White, moderately insoluble $PbCl_2$. Black PbS obtained from acid solutions.

(b) *Identification Test.* Lead sulfate can be dissolved in ammonium acetate or sodium hydroxide. Lead chloride is soluble in hot water. The addition of potassium chromate to either solution precipitates yellow lead chromate:

$$PbSO_4(s) + 3OAc^- \rightarrow Pb(OAc)_3^- + SO_4^{--}$$

$$PbSO_4(s) + 3OH^- \rightarrow Pb(OH)_3^- + SO_4^{--}$$

$$Pb(OAc)_3^- + CrO_4^{--} \rightarrow PbCrO_4(s) + 3OAc^-$$

(c) *Sources of Confusion.* Barium, strontium, and calcium also form white, insoluble sulfates. but they are not soluble in either NH_4OAc or NaOH. Silver and mercury(I) chlorides, also white and insoluble, do not dissolve in hot water. Aluminum, tin, and zinc also give hydroxides that are white and soluble in excess NaOH, but their sulfides are not black.

Magnesium(II) Ion

(a) *Preliminary Indications.* White, gelatinous $Mg(OH)_2$, insoluble in excess base.

(b) *Identification Test.* To a drop of solution under test, add a drop of NH_4Cl solution and then 6 M NH_3 until it is alkaline. No precipitate should form, because the hydroxide ion concentration is kept low by the presence of the ammonium salt (Sec. 2.7). Then add a few drops of Na_2HPO_4 solution. Stir, and rub the inner walls of the tube. The white, crystalline precipitate $MgNH_4PO_4 \cdot 6H_2O$ often forms slowly. If further confirmation is required, see Outline 7, Chapter 11.

(c) *Sources of Confusion.* Magnesium hydroxide can be confused with white, gelatinous hydroxides of lead, cadmium, tin, aluminum, or zinc. Unlike them, it does not dissolve in either NaOH or NH_3, and its precipitation is inhibited by ammonium chloride.

Manganese(II) Ion

(a) *Preliminary Indications.* Pale pink hydrated salts; dilute aqueous solutions virtually colorless; anhydrous compounds white. White $Mn(OH)_2$, insoluble in NaOH and NH_3, darkens on standing [Sec. 6.4(a)]. Pinkish white to pale orange MnS, precipitated only from basic solutions.

(b) *Identification Test.* Test a drop of the solution of the sample for chloride by acidifying it with HNO_3 and adding silver nitrate solution. If no chloride is present, add a drop of 16 M HNO_3 and one-quarter of a spatula-full of sodium bismuthate. Centrifuge down excess bismuthate to see the purple color of MnO_4^-:

$$4Mn^{++} + 5Bi_2O_5 + 18H^+ \rightarrow 4MnO_4^- + 10Bi^{3+} + 9H_2O$$

If a white precipitate was obtained in the chloride test, add $AgNO_3$ solution until no more AgCl forms. Remove the precipitate, acidify the solution with nitric acid, and add bismuthate.

(c) *Source of Confusion.* If the purple color of permanganate fails to form or disappears as fast as it appears, chloride is probably present.

Mercury(I) and Mercury(II) Ions

(a) *Preliminary Indications.* Mercury(I) chloride is insoluble in water; mercury(II) chloride is fairly soluble. The yellow or orange oxide HgO is insoluble in water and alkalies:

$$Hg^{++} + 2OH^- \rightarrow HgO(s) + H_2O$$

$$Hg_2^{++} + 2OH^- \rightarrow Hg(l) + HgO(s) + H_2O$$

The sulfide HgS occurs in red and black forms.

(b) *Identification Test.* Mercury(I) chloride is precipitated when HCl is added to a solution of a mercury(I) salt. It is insoluble in hot water, and turns black when treated with ammonia:

$$2NH_3 + Hg_2Cl_2(s) \rightarrow Hg(l) + HgNH_2Cl(s) + NH_4^+ + Cl^-$$

Mercury(II) chloride is reduced by tin(II) chloride to white Hg_2Cl_2 or to a grey to black mixture of this with mercury:

$$2HgCl_2 + Sn^{++} + 4Cl^- \rightarrow SnCl_6^{--} + Hg_2Cl_2(s)$$

$$Hg_2Cl_2(s) + Sn^{++} + 4Cl^- \rightarrow SnCl_6^{--} + 2Hg(l)$$

(c) *Sources of Confusion.* Of the other two common white insoluble chlorides, $PbCl_2$ dissolves in hot water and AgCl dissolves in ammonia. Reduction of Hg(II) by tin(II) is inhibited if the concentration of chloride is too high.

Nickel(II) Ion

(a) *Preliminary Indications.* Green hydrated salts and solutions; anhydrous salts usually yellow except $NiBr_2$, dark brown, and NiI_2 and NiS, black. Light green hydroxide, insoluble in NaOH, soluble in NH_3 to give blue $Ni(NH_3)_6^{++}$. Its sulfide is precipitated from basic or very weakly acidic solutions.

(b) *Identification Test.* To one drop of the solution under test, add a few drops of water and 15 M NH_3. The solution should be alkaline and blue, without suspended precipitate. Add a drop of dimethylglyoxime solution (Sec. 9.5). A deep red precipitate of nickel dimethylglyoxime indicates the presence of nickel:

$$Ni(NH_3)_6^{++} + 2C_4H_7N_2O_2^- \rightarrow Ni(C_4H_7N_2O_2)_2(s) + 6NH_3$$

(c) *Sources of Confusion.* Other metal ions may give insoluble hydroxides or stable complex ions with ammonia, but only Ni(II) gives the characteristic scarlet complex.

Potassium(I) Ion

(a) *Preliminary Indications.* Fleeting violet flame test, red through cobalt blue glass. No precipitate with NH_3, NaOH, HCl, H_2SO_4, or thioacetamide.

(b) *Identification Test.* To a drop of the solution under test, which should be neutral or weakly acidic, add one drop of 6 M HOAc and a few drops of

$Na_3Co(NO_2)_6$ solution. A yellow precipitate indicates the presence of potassium:

$$2K^+ + Na^+ + Co(NO_2)_6{}^{3-} \rightarrow K_2NaCo(NO_2)_6(s)$$

For a microscopic test see Sec. 6.6.

(c) *Sources of Confusion.* The flame color is not lasting and is easily obscured by the intense yellow of even traces of sodium; cobalt glass blocks the yellow. The cobaltinitrite reagent is unstable; it should be deep reddish amber—discard pale yellow or pink samples. The reagent is also decomposed by strong acid or base. Ammonium salts give a yellow precipitate with cobaltinitrite. Prove that ammonium ion is absent before reporting potassium. The ammonium compound, $(NH_4)_2NaCo(NO_2)_6$, decomposes when the tube containing it is heated in boiling water: evolution of gas is visible and the color of the solution fades to yellow or even pink.

Silver(I) Ion

(a) *Preliminary Indications.* White chloride, insoluble in water, soluble in dilute ammonia to give $Ag(NH_3)_2{}^+$. Brown oxide precipitated by NaOH

$$2Ag^+ + 2OH^- \rightarrow Ag_2O(s) + H_2O$$

is insoluble in excess base, soluble in NH_3. Black sulfide, Ag_2S, precipitated from acid solution.

(b) *Identification Test.* Make the chloride and see if it is soluble in 6 M NH_3. When the solution of the complex is acidified with HNO_3, AgCl should reprecipitate:

$$Ag(NH_3)_2{}^+ + Cl^- + 2H^+ \rightarrow AgCl(s) + 2NH_4{}^+$$

(c) *Sources of Confusion.* Lead(II) chloride is soluble in hot water, and mercury(I) chloride is blackened by ammonia, not dissolved. Failure to get AgCl back when the solution of the complex is acidified is frequently caused by the use of insufficient acid. Check with litmus (T-12).

Sodium(I) Ion

(a) *Preliminary Indications.* Strong, persistent yellow flame test. No precipitate with NH_3, NaOH, H_2SO_4, HCl, or thioacetamide.

(b) *Identification Test.* To a drop of solution under test, which must not be strongly acidic, add 6 drops of magnesium uranyl acetate reagent. Stir, and rub the inner wall of the tube with the rod. Let the tube stand for 5 minutes, stirring occasionally to overcome supersaturation. Formation of pale green yellow crystalline precipitate $NaMg(UO_2)_3(OAc)_9 \cdot 9H_2O$ indicates presence of sodium. A microscopic test is given in Sec. 6.6.

(c) *Sources of Confusion.* The flame test is, if anything, too sensitive; mere traces of sodium will impart a brief yellow color to the flame. Your test should be more persistent. For comparison, dilute one drop of the standard Na^+ solution with 1 ml of 6 M HCl and 4 ml of water in a small beaker or flask and try the test on it.

The sodium magnesium uranyl acetate precipitate is fairly soluble. If you fail to get it, despite a positive flame test, try evaporating the solution of the sample to half or one-quarter of its volume (T-7); use one drop of this more concentrated solution to repeat the test. High concentrations of K^+ ion give a yellow precipitate with the reagent that is paler than the sodium precipitate. This requires 50 mg of K^+ per ml, far more than you should have in the solution you prepared. Other ions: Ag^+, Hg^{++}, Sr^{++}, Sb(III), Sn(IV), and Bi^{3+}, also react with the magnesium uranyl acetate reagent.

Strontium(II) Ion

(*a*) *Preliminary Indications.* Carmine flame test. White sulfate, insoluble in water, NH_4OAc, or NaOH.

(*b*) *Identification Test.* No further tests should usually be required. A microscopic differentiation between strontium and calcium sulfates is given in Sec. 6.6.

(*c*) *Sources of Confusion.* Lead, barium, and calcium ions also give white, insoluble sulfates. Lead sulfate is soluble in NH_4OAc or NaOH (see identification of lead(II) ion in Sec. 6.5). The flame tests should help to distinguish the other two cations from Sr^{++}. Try them on known samples of strontium and calcium salts, and then compare with the result for your sample.

Tin(II) and Tin(IV) Ions

(*a*) *Preliminary Indications.* The white hydroxide $Sn(OH)_2$ and oxide SnO_2 are soluble in NaOH but not in NH_3:

$$Sn(OH)_2(s) + OH^- \rightarrow Sn(OH)_3{}^-$$

$$SnO_2(s) + 2H_2O + OH^- \rightarrow Sn(OH)_6{}^{--}$$

Brown SnS and yellow SnS_2, precipitated from acid solution, soluble in dilute HCl and KOH (SnS less than SnS_2).

(*b*) *Identification Test.* The preliminary indications will usually suffice. The following flame test, though not very sensitive, can be tried for confirmation. Mix in a crucible or small beaker one drop of the solution under test with 10 drops of 12 M HCl. Add a few granules of zinc metal. Dip the end of a test tube filled with cold water in the mixture and then bring the end into the inner cone (reducing zone) of a gas flame. A blue luminescence on the outer wall of the tube indicates the presence of tin.

(*c*) *Causes of Confusion.* Yellow SnS_2, unlike CdS, is soluble in 0.5 M KOH. Unlike yellow As_2S_3, SnS_2 is soluble in 6 M HCl and not in NH_3.

Zinc(II) Ion

(*a*) *Preliminary Indications.* White $Zn(OH)_2$, soluble in both NH_3 and NaOH:

$$Zn(OH)_2(s) + 2OH^- \rightarrow Zn(OH)_4{}^{--}$$

$$Zn(OH)_2(s) + 4NH_3 \rightarrow Zn(NH_3)_4{}^{++} + 2OH^-$$

White ZnS, precipitated from basic or weakly acidic solutions.

(b) *Identification Test.* To one drop of the solution under test, add 6 M NaOH until the solution is basic (T-12) and then add one or two drops in excess. Touch a pipet to this mixture without squeezing the bulb. Some of the solution will rise in the tip by capillary action. Bring the tip down vertically on the center of a square of dithizone paper. A purple red spot indicates the presence of zinc (Sec. 9.5):

$$Zn(OH)_4^{--} + 2C_{13}H_{11}N_4S^- \rightarrow Zn(C_{13}H_{11}N_4S)_2 + 4OH^-$$

(c) *Sources of Confusion.* An orange ring on the paper is caused by NaOH alone. Mercury(I) and (II) and lead(II) give purplish blue spots, and tin(II) gives a pink one. Touch the pipet to some water, and bring the tip down on the center of the spot. The zinc spot will spread as the water flows out; the other spots will not.

6.6   Some Applications of Chemical Microscopy

A microscope with a magnification of 50–100X is satisfactory. If there are three objectives, use the lowest power first and then the medium power; the high-power objective is not used. If you are unfamiliar with the operation of a microscope, ask your instructor for help. Experiment with the illuminator, using the substage condenser, to get the best light for contrast between crystals and background.

The basic technique of precipitation on a microscope slide is carried out in the following way. Place a drop of the solution under test on the slide; it should be about 5–7 mm wide and 1 mm deep—suck off the excess. Place a similar drop of the reagent a few millimeters away from it on the slide. Draw a thin rod from one drop to the other, and tilt the slide so that reagent flows into the solution under test.

Experiment 1. Differentiation of Ca^{++} *from* Ba^{++} *and* Sr^{++}. Clean three slides and spot them with a drop of the standard solutions of the three cations. Use 6 M H_2SO_4 as the precipitating agent. Observe the character of the precipitates through the microscope immediately after precipitation and after 10 minutes of standing. Crystals of calcium sulfate form slowly. Sketch representative crystals in your notebook and describe in words the differences in form. Can you distinguish between barium and strontium by this test?

Experiment 2. Identification of Metal Ions with Potassium Tetrathiocyanato-mercurate(II). Try the reagent first with a small drop of Zn^{++} test solution using the basic precipitation technique. Sketch and describe the crystals of $ZnHg(SCN)_4$. Now dilute a drop of the test solution with 10 drops of water and repeat the test on a drop of this solution. If the crystals differ from those previously obtained, try further dilutions. Test the action of the reagent on one drop each of the standard ion solutions of Cd^{++}, Cu^{++}, Co^{++}, and Pb^{++}. Describe and draw representative crystals.

Experiment 3. Differentiation of Na^+ *from* K^+. Place a small drop of Na^+ standard solution on a microscope slide and evaporate it over a microflame (T-6) or under an infrared lamp. As the slide cools, prepare a slide of K^+ in the same way. Spot each slide, when cool, with a small drop of the reagent, a saturated solution of uranyl acetate (*not* the magnesium uranyl solution in your kit).Induce precipitation of the double salts $NaUO_2(OAc)_3$ and $KUO_2(OAc)_3$ by drawing some of the reagent across the solid spot with a thin rod. The crystals are slow to form. Sketch representative forms in your notebook. The potassium compound is the more soluble of the two, and the test is less sensitive for this ion.

EXERCISES

6.1. Write essential ionic equations for the following reactions in dilute aqueous solution. Any of these could be encountered in the detection of the cation.
   (a) Lead(II) nitrate and aqueous ammonia
   (b) Strontium chloride and sulfuric acid
   (c) Cadmium sulfide and hydrochloric acid
   (d) Bismuth(III) chloride and water [one product is BiOCl(s)]
   (e) Aluminum hydroxide and sodium hydroxide
   (f) Mercury(II) nitrate and hydrogen sulfide
   (g) Iron(III) sulfate and aqueous ammonia
   (h) Cobalt(II) chloride and aqueous ammonia
   (i) Arsenic trichloride [molecular] and hydrogen sulfide
   (j) Copper(II) chloride and aqueous ammonia
   (k) Manganese(II) sulfate and hydrogen sulfide in basic solution
   (l) Ammonium chloride and sodium hydroxide
   (m) Lead(II) hydroxide and sodium hydroxide
   (n) Potassium dichromate and sodium hydroxide
   (o) Mercury(II) nitrate and sodium hydroxide
   (p) Cobalt(II) sulfate and sodium hydroxide
   (q) Antimony(III) sulfide and hydrochloric acid ($SbCl_4{}^-$ is one product)
   (r) Tin(II) chloride and sodium hydroxide in excess
   (s) Iron(III) nitrate and ammonium thiocyanate
   (t) Silver(I) nitrate and aqueous ammonia
   (u) Magnesium sulfate and sodium hydroxide
   (v) Zinc(II) iodide and aqueous ammonia
   (w) Mercury(I) nitrate and hydrochloric acid
   (x) Nickel(II) chloride and aqueous ammonia.

6.2. Separate into partial equations, balance, and recombine to obtain a complete equation. The basic or acidic character of the medium can be inferred from the formulas.
   (a) $CuS(s) + NO_3{}^- \rightarrow NO(g) + Cu^{++} + S(s)$
   (b) $ClO_3{}^- + Cr^{3+} \rightarrow ClO_2(g) + Cr_2O_7{}^{--}$
   (c) $Sn(OH)_3{}^- + Bi(OH)_3(s) \rightarrow Bi(s) + Sn(OH)_6{}^{--}$
   (d) $Mn^{++} + Bi_2O_5(s) \rightarrow MnO_4{}^- + Bi^{3+}$
   (e) $As_2S_3(s) + HO_2{}^- \rightarrow AsO_4{}^{3-} + OH^-$
   (f) $Fe^{++} + O_2 \rightarrow Fe^{3+} + H_2O$
   (g) $Hg_2Cl_2(s) + NH_3 \rightarrow Hg(l) + HgNH_2Cl(s) + NH_4{}^+ + Cl^-$
   (h) $HgCl_2 + Sn^{++} \rightarrow SnCl_6{}^{--} + Hg_2Cl_2(s)$ (HCl solution)
   (i) $Zn(s) + SnCl_6{}^{--} \rightarrow Sn^{++} + Zn^{++}$

Systematic Analysis

of Mixtures

Few reagents are specific for a particular ion. Most are general reagents that react with a large number of other ions. Before identifying an ion, it is usually necessary to separate it from interfering ions. An elaborate scheme has been worked out for metals whereby the cations are first separated into small groups. Then the ions of each group are separated further. More and more specific properties of the ions are used until some final precipitation or color reaction is obtained for each ion that is present. The first group to be precipitated consists of Ag^+, Hg_2^{++}, and Pb^{++}, the common ions that form insoluble chlorides. The *group reagent* that is used to precipitate them is hydrochloric acid. The chlorides are removed and another group is precipitated. The sequence of separations used to divide the 23 metals into five groups can be summarized most conveniently in a block outline.

No system of analysis is a sure solution to all analytical problems. Many elements are not included in the traditional set of metals selected for introductory work, even though some of them, such as titanium, are very common. Nor is an elaborate procedure always necessary or efficient. Modifications and short cuts are appropriate for special samples. The scheme of analysis, in short, is not a substitute for thought. Any analytical procedure must be used intelligently and resourcefully.

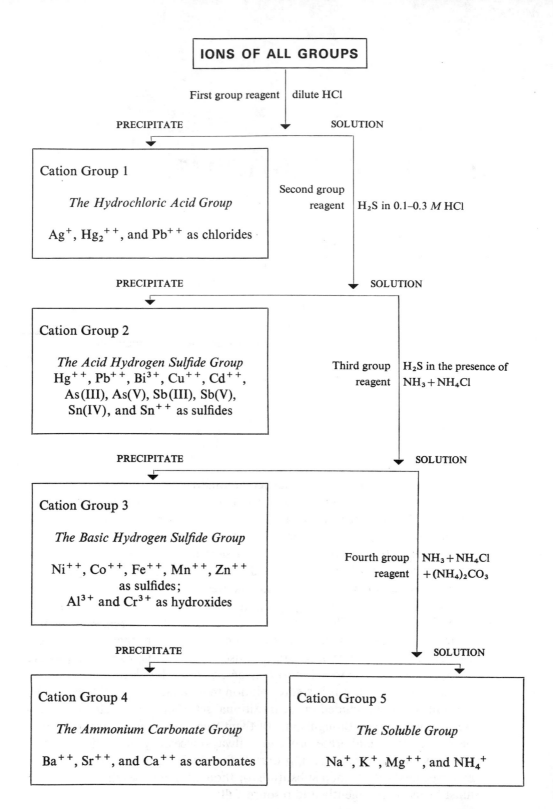

IONS OF ALL GROUPS

First group reagent | dilute HCl

PRECIPITATE   SOLUTION

Cation Group 1

The Hydrochloric Acid Group

Ag^+, Hg_2^{++}, and Pb^{++} as chlorides

Second group reagent | H_2S in 0.1–0.3 M HCl

PRECIPITATE   SOLUTION

Cation Group 2

The Acid Hydrogen Sulfide Group
Hg^{++}, Pb^{++}, Bi^{3+}, Cu^{++}, Cd^{++},
As(III), As(V), Sb(III), Sb(V),
Sn(IV), and Sn^{++} as sulfides

Third group reagent | H_2S in the presence of $NH_3 + NH_4Cl$

PRECIPITATE   SOLUTION

Cation Group 3

The Basic Hydrogen Sulfide Group

Ni^{++}, Co^{++}, Fe^{++}, Mn^{++}, Zn^{++}
as sulfides;
Al^{3+} and Cr^{3+} as hydroxides

Fourth group reagent | $NH_3 + NH_4Cl + (NH_4)_2CO_3$

PRECIPITATE   SOLUTION

Cation Group 4

The Ammonium Carbonate Group

Ba^{++}, Sr^{++}, and Ca^{++} as carbonates

Cation Group 5

The Soluble Group

Na^+, K^+, Mg^{++}, and NH_4^+

In the chapters that follow, the analytical procedures are given largely in outline form. All outlines start with the treatment of a sample, and the directions for this up to centrifugation or division of the sample form a rectangle. An arrow leads from this to separate rectangles in which subsequent treatment of the precipitate and solution is described. The succession of rectangles continues until all ions have been separated and identified. The rectangles in which final identifications are described are set on the right-hand side or at the bottom of the outline and are shaded to distinguish them from the rest. To conserve space and keep the organization of the analysis clearly in view, the directions are as condensed as possible. Explanations, precautions, and special instructions are relegated to the notes that follow each outline; these are referred to by number in the outline. Common techniques have been described in Sec. 3.3, and references are made to these; for example, T-8 refers to Technique 8.

SEVEN

Cation Group 1.

The Hydrochloric Acid Group:

Ag^+, Hg_2^{++}, Pb^{++}

7.1   Properties of the Ions

The ions of this analytical group are precipitated as insoluble chlorides by addition of a slight excess of hydrochloric acid. The positions of these elements and of the less common members of the group in the periodic table are shown below. Although they resemble each other in forming insoluble chlorides, the ions in other reactions show an interesting diversity of behavior. Thus, once the group has been precipitated, it is easy to separate and identify the ions by using their more specific characteristics.

IB	IIB	IIIA	IVA
(Cu^I)			
Ag^I			
(Au^I)	Hg^I	(Tl^I)	Pb^{II}

Some characteristics of the ions of this group are given in Table 7.1. The simple ions Ag^+ and Pb^{++} have stable electronic configurations that are not excited by the low-energy radiation of visible light; the ions and such compounds as AgCl

129

and $Pb(NO_3)_2$ are therefore colorless. The monatomic mercury(I) ion Hg^+, if it existed, would have an unpaired $6s$ valence electron. A more stable ion is obtained by pairing the valence electrons of two such ions to give $^+Hg:Hg^+$ or Hg_2^{++}. This stable electron configuration likewise corresponds to a colorless ion and colorless compounds such as Hg_2Cl_2.

Table 7.1

Some Characteristics of the Ions of Cation Group 1

Ion	Electronic configuration	Ionic radius, in Å	$\dfrac{Charge}{Radius}$
Ag^+	(Kr core) $4d^{10}$	1.26	0.79
Pb^{++}	(Xe core) $4f^{14}5d^{10}6s^2$	1.21	1.65
$[Hg^+]^*$	(Xe core) $4f^{14}5d^{10}6s^1$	~1.3	~0.75
Hg_2^{++}	$^+Hg:Hg^+$	—	—

* Hypothetical, not stable in this form.

The interaction of these ions with other ions or with molecules is not entirely electrostatic. The electrostatic effect of an ion depends directly on its charge and inversely on its radius. The charge/radius ratio of Ag^+ ion, for example, is not very different from that of potassium ion (0.75), leading us to expect that compounds of both ions would have high solubilities and little tendency to form complexes. This is manifestly not in accord with the facts. Potassium chloride may be soluble in water, but silver chloride is not. Moreover, silver ion forms a number of well-known complexes such as $Ag(NH_3)_2^+$, $AgCl_2^-$, and $Ag(S_2O_3)_2^{3-}$. We must therefore suppose some interaction between Ag^+ ion and NH_3, Cl^-, or $S_2O_3^{--}$ that is nonelectrovalent. It is most simply described as partial covalent bonding: as the ligands and Ag^+ are brought together, partial overlapping of their electron clouds occurs. Such nonelectrovalent interactions are characteristic of ions with outer-shell configurations of d^{10} or $d^{10}s^2$.

The Pb^{++} ion is only slightly larger than Ag^+ ion, yet it has twice the charge, so that its electrostatic effect is twice as large. Despite this, lead(II) chloride is the most soluble of the three chlorides of this group. That is because the strong electrostatic effect of Pb^{++} ion is exerted not only on Cl^- ions but also on water molecules. In order for precipitation to occur, the water molecules in the hydration shell of the cation must be replaced by anions:

$$Pb(H_2O)_4^{++} + 2Cl^- \rightarrow PbCl_2(s) + 4H_2O$$

The extent of this reaction depends on the competitive action of H_2O and Cl^- on

Pb^{++}. Its double charge makes Pb^{++} hold H_2O molecules more tightly than Ag^+ does, and formation of $PbCl_2$ occurs to only a limited extent.

The solubilities of the halides of the three ions of this group always fall in the following order:

fluoride (most soluble) > chloride > bromide > iodide

This is again evidence of interaction beyond electrostatic. If charge and radius were the only ionic properties that counted, crystals with the smallest anion (fluoride) would be the most tightly bound together and the least soluble. The opposite is true. The largest anion (iodide) has the most diffuse electronic charge cloud and is best capable of sharing its electrons with a cation. The change of electronic structure that occurs when such anions combine with a cation of this group is also manifested in absorption of visible light. Colorless silver and iodide ions combine to give yellow silver iodide. Mercury(I) iodide, lead(II) iodide, and silver bromide are also yellow, although the last of these is pale.

7.2 Precipitation of the Group

Silver and mercury(I) chlorides have very low solubilities and can be precipitated almost completely by addition of a slight excess of hydrochloric acid. Lead(II) chloride is much more soluble and will not precipitate at all, unless the concentration of Pb^{++} ion is high. Even at best, enough lead is left in solution to give a precipitate of lead sulfide with Cation Group 2.

Hydrochloric acid, rather than ammonium chloride or another soluble chloride, is used to precipitate the group because it supplies hydrogen ions as well as chloride ions. The solution must be sufficiently acid to prevent precipitation of the oxychlorides of bismuth(III) and antimony(III):

$$Bi^{3+} + Cl^- + H_2O \leftrightharpoons BiOCl(s) + 2H^+$$

A slight excess of hydrochloric acid is desirable in order to keep the concentrations of the three cations left in solution as low as possible; for example,

$$[Ag^+][Cl^-] = K_{sp}$$

A large excess is avoided, lest the formation of weakly dissociated chloro complexes

$$Pb^{++} + 3Cl^- \leftrightharpoons PbCl_3^-$$

$$Ag^+ + 2Cl^- \leftrightharpoons AgCl_2^-$$

$$Hg_2^{++} + 4Cl^- \leftrightharpoons HgCl_4^{--} + Hg(l)$$

should so reduce the concentration of cations that precipitation would be incomplete or not occur at all:

$$[Pb^{++}][Cl^-]^2 < K_{sp}$$

7.3   Separation and Identification of the Ions

Precipitation of the group made use of a common property of the three cations—formation of an insoluble chloride. The ions otherwise differ from each other in so many ways that it is easy to separate them. All the chlorides become more soluble as the temperature is raised, but the effect is noticeable only with lead chloride. At 100°C, 1 ml of water will dissolve ten times as much silver chloride as it does at 25°C, but this is still only 0.021 mg of AgCl. The solubility of lead chloride only triples with this increase in temperature, but 1 ml of water dissolves over 30 mg of this more soluble precipitate at 100°C:

$$PbCl_2(s) \xrightarrow{\Delta} Pb^{++}(aq) + 2Cl^-(aq)$$

Lead chloride in the hot water extract or even as a precipitate in the cold is readily converted to less soluble, bright yellow lead chromate:

$$PbCl_2(s) + CrO_4^{--} \leftrightharpoons PbCrO_4(s) + 2Cl^- \qquad K = 8 \times 10^{10} \ M$$

The residue from the hot water extraction may be silver or mercury(I) chloride or a mixture of the two. These two solids differ in their reaction with ammonia, so that this reagent can be used to separate them. Silver chloride is dissolved to give the weakly dissociated ammonia complex of silver(I):

$$AgCl(s) + 2NH_3 \rightarrow Ag(NH_3)_2^+ + Cl^-$$

When the solution of the complex is acidified with nitric acid, ammonia is converted to ammonium ion, which does not react with silver ion, and silver chloride reprecipitates:

$$Ag(NH_3)_2^+ \leftrightharpoons Ag^+ + 2NH_3$$

$$2NH_3 + 2H^+ \rightarrow 2NH_4^+$$

$$Ag^+ + Cl^- \rightarrow AgCl(s)$$

Net: $Ag(NH_3)_2^+ + Cl^- + 2H^+ \rightarrow AgCl(s) + 2NH_4^+$

Ammonia causes mercury(I) chloride to disproportionate. Half the mercury(I) is reduced to free mercury, which is black in finely divided form:

$$2e^- + Hg_2Cl_2(s) \rightarrow 2Hg + 2Cl^-$$

and half is oxidized to mercury(II) amidochloride:

$$4NH_3 + Hg_2Cl_2(s) \rightarrow 2HgNH_2Cl(s) + 2NH_4^+ + 2e^-$$

(A complete equation can be derived by combining these partial equations.) The formation of the black residue is usually sufficient indication of the original presence of mercury(I).

When a large amount of mercury is present, small amounts of silver ion can be lost by reduction of silver chloride to metallic silver:

$$2AgCl(s) + Hg(l) \rightarrow Hg_2Cl_2(s) + 2Ag(s)$$

If the silver test is indecisive or negative, silver must be sought in the black residue from the ammonia treatment. Aqua regia, a mixture of concentrated nitric and hydrochloric acids, dissolves the residue. Silver is converted to a precipitate of silver chloride

$$Ag + Cl^- \rightarrow AgCl(s) + e^-$$

$$e^- + NO_3^- + 2H^+ \rightarrow NO_2(g) + H_2O$$

$$\overline{NO_3^- + Ag + Cl^- + 2H^+ \rightarrow AgCl(s) + NO_2(g) + H_2O}$$

or, with excess chloride, into a chloro complex

$$AgCl(s) + Cl^- \rightarrow AgCl_2^-$$

The excess nitric and hydrochloric acids are removed by evaporation and reaction. The equation for the latter can be derived from the skeleton,

$$Cl^- + NO_3^- \rightarrow NOCl(g) + Cl_2(g),$$

by separating it into partial equations and balancing them. When the residual solution is diluted with water, the dissociation of $AgCl_2^-$ increases and silver chloride reprecipitates:

$$AgCl_2^- \rightarrow AgCl(s) + Cl^-$$

Aqua regia also dissolves the mercury part of the black residue. Equations can be derived from the skeletons:

$$NO_3^- + Hg(l) + Cl^- \rightarrow HgCl_4^{--} + NO_2(g)$$

and

$$NO_3^- + HgNH_2Cl(s) + Cl^- \rightarrow HgCl_4^{--} + N_2(g) + NO_2(g)$$

The tetrachloromercurate(II) ion, $HgCl_4^{--}$, is formed in both reactions. If confirmation of the presence of mercury is desired, it can be obtained, after removal of silver chloride from the mixture, by adding tin(II) chloride to the solution. Mercury is reduced first to white Hg_2Cl_2:

$$2HgCl_4^{--} + Sn^{++} \rightarrow SnCl_6^{--} + Hg_2Cl_2(s)$$

and then to black mercury:

$$Hg_2Cl_2(s) + 4Cl^- + Sn^{++} \rightarrow 2Hg(l) + SnCl_6^{--}$$

7.4 Illustrative Numerical Calculations

(a) The Relation Between Solubility and Solubility Product

The equilibrium between mercury(I) chloride and its saturated solution might be expected to be represented by

$$Hg_2Cl_2(s) \leftrightharpoons Hg_2^{++} + 2Cl^- \qquad K_{sp} = [Hg_2^{++}][Cl^-]^2 = 1.3 \times 10^{-18} \, M^3$$

Put S equal to the molar solubility of Hg_2Cl_2, that is, the number of moles per liter of the salt in the saturated solution. Then according to the equation, one Hg_2Cl_2 unit gives one Hg_2^{++} ion and 2 Cl^- ions. Thus we have $[Hg_2^{++}] = S$ and $[Cl^-] = 2S$, provided the above equilibrium is the only one of consequence. We therefore have to solve

$$1.3 \times 10^{-18} \ M^3 = S(2S)^2$$

and we obtain $S = 6.9 \times 10^{-7} \ M$.

In precipitation of Group 1 practice solution, one drop of 6 M HCl supplies chloride for the precipitate and leaves excess chloride at approximately 0.35 M concentration. The solubility S' is now measured by $[Hg_2^{++}]$; the total chloride comes partly from HCl and partly from Hg_2Cl_2. We now have

$$S'(0.35 + 2S')^2 \simeq S'(0.35)^2 = 1.3 \times 10^{-18} \ M^3$$

whence $S' = 1.3 \times 10^{-17} \ M$. We thus predict that the excess chloride will cause a very large decrease in solubility.

Commentary   Because of various assumptions, implicit and explicit, we cannot expect the numerical answers to be more than rough approximations. The measured solubility in pure water is about eleven times as large as we calculated, because other reactions occur that we did not take into account. Mercury(I) ions hydrolyze and also disproportionate, giving species such as Hg_2OH^+, $Hg(OH)_2$, $HgCl^+$, and $HgCl_2$. The molar solubility is then no longer equal to $[Hg_2^{++}]$ but has to be expressed in terms of the total mercury content of the solution. Excess hydrochloric acid will suppress hydrolysis but encourage disproportionation.

Several responses can be made to the natural query, Are such inaccurate calculations worthwhile? First of all, the practical difference between a measured solubility of 0.0000075 M and a calculated value of 0.00000069 M is not very significant; they are both very small. There is nothing wrong with our calculations, aside from a minor error due to neglect of interionic forces, given the premise that the only equilibrium is the one between Hg_2Cl_2 and Hg_2^{++} and Cl^- ions. The algebra and arithmetic are then comparatively simple, but as we gain experience in such calculations it becomes easy to extend them to include the other equilibria. The comparison of calculated and observed solubilities does serve to remind us of the limitations of too simple a view of electrolytic solutions. The simple view takes in the highlights—and novices should concentrate on them first—but lacks the fascinating and baffling details of the complete picture.

(b) Solubility and Complex Formation

The solubility of silver chloride is increased by the addition of a large excess of either hydrochloric acid or ammonia. At low concentrations of chloride, the principal complex is $AgCl_2^-$ and the equilibrium is represented by

$$AgCl(s) + Cl^- \leftrightarrows AgCl_2^- \qquad K = \frac{[AgCl_2^-]}{[Cl^-]} = 2.0 \times 10^{-5}$$

Suppose that silver chloride is shaken with 0.10 M HCl. For simplicity, we assume that negligible concentrations of chloride are contributed by AgCl or consumed to form $AgCl_2^-$. This makes $[Cl^-] = 0.10$ and $[AgCl_2^-] = 0.10 \times 2 \times 10^{-5} = 2.0 \times 10^{-6}$ M. We expect $[Ag^+]$ to be low because of the common-ion effect. If there are no other forms of silver in solution, 2.0×10^{-6} M is then the solubility of silver chloride.

Commentary The measured solubility is 3.0×10^{-6} M. We can get a better calculated value by allowing for the presence of dissolved silver chloride ion pairs AgCl(aq) and for higher complexes such as $AgCl_3^{--}$. The solubility is the total concentration of silver in solution

$$S = [Ag^+] + [AgCl] + [AgCl_2^-] + [AgCl_3^{--}] + [AgCl_4^{3-}] \qquad (7.1)$$

We define the following constants:

$$K_{s1} = [AgCl] \qquad K_{s2} = \frac{[AgCl_2^-]}{[Cl^-]}$$

$$K_{s3} = \frac{[AgCl_3^{--}]}{[Cl^-]^2} \qquad K_{s4} = \frac{[AgCl_4^{3-}]}{[Cl^-]^3} \qquad (7.2)$$

By combining equations (7.1) and (7.2), you can show that

$$S = \frac{K_{sp}}{[Cl^-]} + K_{s1} + K_{s2}[Cl^-] + K_{s3}[Cl^-]^2 + K_{s4}[Cl^-]^3 \qquad (7.3)$$

Introducing numerical values for the K's and $[Cl^-]$, we have

$$S = \frac{1.8 \times 10^{-10}}{10^{-1}} + 2.5 \times 10^{-7} + 2.0 \times 10^{-5} \times 10^{-1} + 3.0 \times 10^{-5} \times 10^{-2}$$
$$+ 2.0 \times 10^{-4} \times 10^{-3}$$

whence $S = 2.8 \times 10^{-6}$ M. This is in good agreement with the observed solubility of 3.0×10^{-6} M. The greater part of the total is due to formation of $AgCl_2^-$. At higher concentrations of HCl, greater contributions are made by $AgCl_3^{--}$ and $AgCl_4^{3-}$.

The solubility of AgCl in aqueous ammonia can be treated similarly. We know the solubility product of the salt

$$AgCl(s) \leftrightharpoons Ag^+ + Cl^- \qquad K_{sp} = [Ag^+][Cl^-] = 1.8 \times 10^{-10} \ M^2$$

and the overall formation constant of the complex

$$Ag^+ + 2NH_3 \leftrightharpoons Ag(NH_3)_2^+ \qquad \beta_2 = \frac{[Ag(NH_3)_2^+]}{[Ag^+][NH_3]^2} = 1.6 \times 10^7 \ M^{-2}$$

Silver ion is common to both chemical equations, and $[Ag^+]$ is a common factor in the two equilibrium quotients. Combining these expressions we get

$$AgCl(s) + 2NH_3 \leftrightharpoons Ag(NH_3)_2^+ + Cl^-$$

$$K = K_{sp}\beta_2 = \frac{[Ag(NH_3)_2^+][Cl^-]}{[NH_3]^2} = 2.9 \times 10^{-3}$$

If virtually all the silver goes into the form of the complex and all the chloride remains free, the molar solubility is given by

$$S = [Cl^-] = [Ag(NH_3)_2{}^+]$$

If we shake silver chloride with 0.50 M ammonia, the formation of the complex consumes $2S$ moles, leaving at equilibrium $[NH_3] = 0.50 - 2S$. The expression to be solved is thus

$$K = \frac{S \times S}{(0.50-S)^2} = 2.9 \times 10^{-3} \quad \text{or} \quad K^{1/2} = \frac{S}{0.50-S} = 5.4 \times 10^{-2}$$

whence $S = 2.6 \times 10^{-2}\ M$. This agrees well with the measured solubility of $2.9 \times 10^{-2}\ M$.

> **Commentary** Silver ion combines with ammonia in steps forming $AgNH_3{}^+$ as well as $Ag(NH_3)_2{}^+$. At all but very low concentrations of ammonia, virtually all the silver goes into the complex with higher ammonia content. At high concentrations of ammonia, further complications arise: the calculated solubility in $5\ M\ NH_3$ is only about half that observed, namely, 0.5 M. Here we have to take account of the effect of the ammonia medium on the equilibrium constants and the formation of other complexes, such as $AgCl(aq)$.

(c) Disproportionation of Mercury(I)

In studying a reaction such as $Hg_2{}^{++} \leftrightharpoons Hg(l) + Hg^{++}$, we have various ways of arriving at its equilibrium constant. One is by direct analysis of the equilibrium mixture prepared by shaking metallic mercury with various solutions of mercury(II) nitrate (Table 7.2).

Table 7.2

Analytical Data for the Disproportionation of Mercury(I) at 25°C

Solution	$[Hg^{++}]$	$[Hg(aq)]$	$[Hg^+]$	$[Hg_2{}^{++}]$	K'	K''
A	0.00216 M	$3.0 \times 10^{-7}\ M$	0.0516 M	0.0258 M	4.1×10^7	4.0×10^8
B	0.00419	3.0×10^{-7}	0.1004	0.0502	8.0×10^7	4.0×10^8
C	0.00461	3.0×10^{-7}	0.1106	0.0553	8.8×10^7	4.0×10^8

If the mercury(I) ion were Hg^+, the equilibrium would be expressed by

$$Hg^{++} + Hg(l) \leftrightharpoons 2Hg^+ \qquad K' = \frac{[Hg^+]^2}{[Hg^{++}][Hg]}$$

but such ratios are not constant, as the values in column 6 of Table 7.2 show. On the other hand, for $Hg^{++} + Hg(l) \leftrightharpoons Hg_2^{++}$, the values of K'' are constant, confirming Hg_2^{++} as the correct formula for the mercury(I) ion.

For the disproportionation reaction, which is the reverse of the formation of Hg_2^{++}, the equilibrium constant will be the reciprocal of K''. Moreover, it is customary to combine K'' with the concentration of mercury in solution because this is fixed by its equilibrium with liquid mercury. Thus we have

$$Hg_2^{++} \leftrightharpoons Hg(l) + Hg^{++} \qquad K = \frac{[Hg^{++}]}{[Hg_2^{++}]} = 8.3 \times 10^{-3}$$

from analytical data. The small value indicates a low extent of reaction.

The disproportionation reaction can also be studied in electric cells. We have for the standard electrode potentials at 25°C

$$2e^- + Hg_2^{++} \rightarrow 2Hg(l) \qquad E^\circ = 0.792 \text{ V}$$

$$2e^- + 2Hg^{++} \rightarrow Hg_2^{++} \qquad E^\circ = 0.908 \text{ V}$$

To get the required net equation, we have to subtract the second half-reaction from the first. Dividing the result by 2 will not affect E°:

$$Hg_2^{++} \leftrightharpoons Hg(l) + Hg^{++} \qquad E^\circ = -0.116 \text{ V}$$

The negative value of E° again indicates a low degree of reaction when the two ions are at approximately unit concentration. The relation between E° and K is given by equation (1.2):

$$\log K = \frac{E^\circ}{k_N/n} = \frac{-0.116}{0.0592/1} = -1.96$$

or $K = 1.10 \times 10^{-2}$.

A third way of finding equilibrium constants uses calorimetric data, standard Gibbs energies of formation. For Hg^{++}, ΔG_f° is 39.38 kcal mol^{-1} at 25°C, and for Hg_2^{++} it is 36.79 kcal mol^{-1}. The value of ΔG_f° for Hg(l) or any other element in its standard state is zero. Thus we have

$$\Delta G^\circ = \Delta G_f^\circ(Hg^{++}) - \Delta G_f^\circ(Hg_2^{++})$$

$$= 39.38 - 36.79 = 2.59 \text{ kcal}$$

The relation between ΔG° and K is given by equation (1.3):

$$\log K = -\frac{\Delta G^\circ}{2.303RT} = -\frac{2.59}{1.364} = -1.90$$

or $K = 1.26 \times 10^{-2}$.

Commentary The power of thermodynamics lies in establishing relationships between seemingly unrelated measurements. Here chemical analysis, electric cells, and calorimetric measurements all yield values of an equilibrium constant. The three values 0.83×10^{-2}, 1.10×10^{-2}, and 1.26×10^{-2} are in fair agreement. Most of the experimental effort goes into establishing the value of the power of 10, the exponent -2; this is the most important part of the number. In any case, these are difficult ions to study precisely.

Complex formation or precipitation has a profound effect on the extent of disproportionation. If chloride is present, for example, it ties up Hg_2^{++} in a precipitate and Hg^{++} in various complexes ranging from $HgCl_2$ to $HgCl_4^{--}$. From cell data we have

$$2e^- + Hg_2Cl_2(s) \rightarrow 2Hg(l) + 2Cl^- \qquad E° = 0.27 \text{ V}$$

$$2e^- + 2HgCl_4^{--} \rightarrow Hg_2Cl_2(s) + 6Cl^- \qquad E° = 0.61 \text{ V}$$

so Hg_2Cl_2 is a poorer oxidizing agent and better reducing agent than Hg_2^{++}. For the overall reaction:

$$Hg_2Cl_2(s) + 2Cl^- \rightarrow Hg(l) + HgCl_4^{--} \qquad E° = 0.27 - 0.61 \text{ V}$$

The value of $E° = -0.34$ V is more negative than that in the absence of chloride, meaning that less disproportionation occurs. When ammonia is added to Hg_2Cl_2, as in the analysis of Group 1, the situation is reversed and virtually all the Hg_2Cl_2 is converted to Hg and $HgNH_2Cl$.

EXPERIMENTAL PART

7.5   Analysis of Known and Unknown Solutions

(1) Use 5 drops of a *known* or *practice* solution that contains about 15 mg of Pb^{++}, 10 mg of Hg_2^{++}, and 2 mg of Ag^+ per ml.

(2) The *unknown* may be a solution containing the ions of this and possibly other groups. If it is basic, acidify with 6 M HCl. If the unknown is a solid, take a representative sample (T-1) and dissolve about 20 mg [T-3(b)] in a few drops of 6 M HNO_3.

The directions for the analysis are given in Outline 1. The superscript numbers in the instructions direct your attention to the notes that follow the outline for explanations, precautions, and further instructions. The symbols T-8, T-12, etc., refer to the techniques described in Sec. 3.3. Consult your instructor for the type of notebook record that should be kept for the analysis.

OUTLINE 1. The Systematic Analysis of Cation Group 1

To a suitable sample (Sec. 7.5) of known or unknown solution in a centrifuge tube, add 1 drop[1] [T-3(a), T-4(a)][2] 6 M HCl and stir.[3] If a ppt forms,[4] allow it to settle and add another drop of HCl to test for completeness of pptn. Continue addn of HCl until no more ppt forms, but avoid a large xs.[5] Stir for 1–2 minutes to give time for slow pptn of $PbCl_2$. Centrifuge, using another centrifuge tube filled with an equal volume of water as a counterbalance (T-8). Withdraw as much as possible of the clear soln with a capillary pipet (T-9). Transfer soln to a test tube and check again for completeness of pptn. Wash the residue (T-10) with a few[6] drops of very dil HCl.[7]

Solution 1.1. Ions of later groups, including Pb^{++}, and H^+ and Cl^-.

Precipitate 1.1. AgCl, Hg_2Cl_2, $PbCl_2$. Add about 5 drops hot, distilled water.[8] Warm in hot water bath and stir occasionally for a few minutes. Centrifuge and separate soln from ppt quickly. Transfer soln to test tube.

Solution 1.2. Pb^{++}, $PbCl^+$, Cl^-. Add 1 drop 6 M HOAc[9] and several[6] drops K_2CrO_4. Yellow ppt, $PbCrO_4$: **presence of Pb.**[10]

Precipitate 1.2. AgCl, Hg_2Cl_2, ($PbCl_2$). If Pb is found in Soln 1.2, repeat extn of residue with hot water until extract gives only a faint turbidity with K_2CrO_4.[11]

To the Pb-free residue, add several drops[6] of 15 M NH_3. Use a rod to disperse caked residue in the reagent. Centrifuge. Transfer soln to a test tube.[12]

Precipitate 1.3. $HgNH_2Cl$, Hg (black), (AgCl, Ag). Black residue is usually sufficient indication of **presence of Hg(I).**

If Ag test is indecisive or negative or if confirmation of Hg(I) is required, treat the black residue with 1 drop 16 M HNO_3 and 3 drops 12 M HCl (aqua regia). Warm in hot water bath until reaction ceases. Transfer mixture to crucible. Evaporate carefully almost to dryness (T-4).[14] Dilute with few drops of water and transfer to a centrifuge tube.

Solution 1.3. $Ag(NH_3)_2^+$, Cl^-. Acidify with 6 M HNO_3. Verify with litmus (T-12).[13] White ppt, AgCl: **presence of Ag.**

Solution 1.4. $HgCl_4^{--}$. Add a few drops $SnCl_2$ soln. White, grey, or black ppt, Hg_2Cl_2, Hg: **presence of Hg(I).**

Precipitate 1.4. AgCl (white). Add a few drops of 6 M NH_3. AgCl should dissolve. Reppt with 6 M HNO_3 (see Soln 1.3). White ppt, AgCl: **presence of Ag.**

Notes on Outline 1

1. Small quantities of reagents are usually required. One drop (0.05 ml) of 6 M HCl contains $6 \times 0.05 = 0.3$ mmol. If a sample contains 20 mg of silver, a very large amount, this is equivalent to 0.19 mmol. Thus, one drop of 6 M HCl supplies more than enough chloride for complete precipitation.

2. Such references are to the techniques described in Sec. 3.3.

3. Thorough mixing is difficult in a centrifuge tube because of the narrow bottom (T-5). Work the glass rod up and down, and rub the inner wall to induce crystallization of $PbCl_2$, which readily forms supersaturated solutions.

4. If no precipitate forms, Ag^+ and Hg_2^{++} ions are definitely absent, but Pb^{++} ion may be at too low a concentration to precipitate with Group 1 (it will then appear in Group 2), or $PbCl_2$ may have supersaturated. Supersaturation is overcome by vigorous stirring, rubbing the inner walls with a rod, and patient waiting. Allow 5 minutes before concluding that Group 1 is absent.

5. A large excess of reagent may dissolve the chlorides.

6. In these directions "a few" is taken to mean 1 or 2 drops and "several" to mean 2 to 4 drops. Learn to judge for yourself from the size of the precipitate how much reagent to add. Use the least amount of reagent that will produce the required result.

7. The residence is washed with very dilute HCl to minimize the loss of $PbCl_2$. Dilute one drop of 6 M HCl with about 10 drops (0.5 ml) of water in a test tube and mix well [T-5(a)].

8. Use water from a tube suspended in the water bath. Start heating it before beginning the analysis so that it will be ready when you need it. Do not use water directly from the water bath; if this has a lead top, enough lead is introduced to give a false test.

9. The acetic acid prevents precipitation of other chromates, such as $CuCrO_4$ or $(BiO)_2CrO_4$, which may appear at this point through failure to wash the group precipitate carefully.

10. Centrifuge to determine the bulk of the precipitate. It is useful to observe the size of tests on known solutions, for these help you to distinguish between normal and trace amounts in unknowns. Traces are generally caused by contamination and should not be reported.

When large amounts of bismuth are present, it may sometimes carry through to this point through careless technique. If this is suspected, test the solubility of the yellow precipitate in 6 M NaOH; $PbCrO_4$ will dissolve, but $(BiO)_2CrO_4$ will not.

11. If lead is not almost completely removed from the precipitate, it can coat the chlorides with insoluble $Pb(OH)_2$ when ammonia is added and thus prevent their reaction with the reagent.

12. A test tube is better for the acidification of Solution 1.3 because of the ease of mixing in such tubes (T-5).

13. This is a more difficult operation than the beginner may realize. Silver is often missed through failure to mix solution and nitric acid well and to get a definite acid reaction. If you see a cloudy white layer floating on top of a clear one, mixing has not been vigorous enough. Note particularly, as discussed in T-12, that a false acid reaction can be obtained if the stirring rod touches an upper part of the tube that is wet with acid. Mix as described in T-5(a).

14. The purpose of evaporation is to remove excess HCl and destroy nitric acid, which would otherwise interfere with the test for mercury. But if evaporation is carried too far, some $HgCl_2$ may be lost. Use a microflame (T-7) and hold the crucible in a wire

loop made from a pipe cleaner [Fig. 3.4(b)]. Wave the crucible back and forth *above*, not in, the flame, withdrawing it momentarily if evaporation becomes too vigorous. *Stop heating when a drop or two of solution remains;* further evaporation will occur as the hot solution stands.

EXERCISES

7.1. For each element in this analytical group, give its position in the periodic table and account for its occurrence in this analytical group.

7.2. Give the electronic configurations of the outer shells of Ag^+, K^+, Ba^{++}, and Pb^{++}.

7.3. Some properties of KCl and AgCl are given below. Account for the similarities and differences.

	KCl	AgCl
Crystal structure type	NaCl	NaCl
Cation radius, Å	1.33	1.26
Melting point, °C	776	455
Lattice energy, kcal mol^{-1}	169	217
Solubility in water at 25°C, M	5	0.000013

7.4. The hydration energies of Pb^{++} and Ba^{++} are, respectively, 353 and 311 kcal mol^{-1}. Account for the difference.

7.5. Copper forms two ions: Cu^+, radius 0.96 Å, and Cu^{++}, radius 0.72 Å. Account for the fact that $CuCl_2$ is soluble in water but CuCl is not.

7.6. Lead bromide is colorless, whereas lead iodide is yellow. Account for the difference. How would you expect them to compare in solubility?

7.7. Thallium(I) ion, Tl^+, forms an insoluble chloride TlCl, whereas bismuth(III) chloride, $BiCl_3$, is soluble (though extensively hydrolyzed). Look up the position of the two elements in the periodic table. How would these ions compare with Pb^{++} in electronic structure? Account for the difference in solubility of their chlorides.

7.8. Give the colors of Ag^+, $Pb(NO_3)_2$, Hg_2SO_4, PbI_2, $Ag(NH_3)_2{}^+$, $Hg_2(NO_3)_2 \cdot 2H_2O$, Hg_2Cl_2, $PbCrO_4$, AgBr, $Hg_2{}^{++}$. Which are (a) colorless because of stable electronic structures, (b) colored because of a colored anion, (c) colored because of a partially covalent bond?

7.9. Classify the following as soluble, sparingly soluble, or almost insoluble in water: Ag_2SO_4, $Hg_2(NO_3)_2$, $Pb(OAc)_2$, AgF, $PbCrO_4$, AgBr, $HgNH_2Cl$, $Ag(NH_3)_2Cl$, Hg_2I_2.

7.10. Explain why a slight excess of hydrochloric acid is used in precipitating the group. Why is a large excess to be avoided?

7.11. Account for the following operations in the analysis by Outline 1:
(a) The addition of acetic acid before the identification of lead.
(b) The continued washing of the group precipitate with hot water.
(c) The addition of aqua regia to the mercury precipitate.

7.12. Devise the briefest possible scheme of analysis for each of the following samples. You may use separations actually employed in the analytical procedure or invoke other properties of the ions.

(a) A solid sample suspected of being mercury(I) nitrate.

(b) A white solid that might be either calomel (Hg_2Cl_2), a drug, or corrosive sublimate ($HgCl_2$), a poison.

(c) The nitric acid solution of a metal suspected of being sterling silver, a silver–copper alloy.

(d) A yellow pigment suspected of being either lead chromate or cadmium sulfide.

7.13. Write equations for the successive reactions of each ion of this group from precipitation of the chloride to final identification.

7.14. The equation $NH_3 + Hg_2Cl_2(s) \rightarrow Hg(l) + HgNH_2Cl(s) + H^+ + Cl^-$ is balanced, and yet it is incorrect. Why are 2 moles of ammonia required?

7.15. Calculate the solubility of silver chloride in mg AgCl per ml under the conditions of the group precipitation: 0.35 M HCl, initially 1.7 mg of Ag^+ per ml.

7.16. Calculate the solubility of silver bromide in 10 M aqueous ammonia, assuming that it dissolves to give only $Ag(NH_3)_2^+$ and Br^- ions. The overall formation constant of the complex is 1.6×10^7 M^{-2}; consult Appendix A.2 for the solubility product of AgBr.

7.17. From the solubility products of lead chloride and lead chromate given in Appendix A.2, calculate the equilibrium constant for

$$PbCl_2(s) + CrO_4^{--} \rightleftharpoons PbCrO_4(s) + 2Cl^-$$

Will any lead chloride form if solid lead chromate is shaken with a solution that is 0.0010 M in chromate and 1.0 M in chloride?

7.18. For the half-reaction $SnCl_6^{--} + 2e^- \rightarrow Sn^{++} + 6Cl^-$, $E°$ is 0.15 V. The standard electrode potentials for reduction of mercury chlorides are given at the end of Sec. 7.4. Calculate $E°$ and K for each of the following reactions:

(a) $2HgCl_4^{--} + Sn^{++} \rightarrow SnCl_6^{--} + Hg_2Cl_2(s)$

(b) $Hg_2Cl_2(s) + 4Cl^- + Sn^{++} \rightarrow SnCl_6^{--} + 2Hg(l)$

7.19. From the following standard Gibbs energies of formation: AgCl(s), -26.22 kcal mol^{-1}; Hg_2Cl_2(s), -50.35; Ag^+, 18.43; Hg_2^{++}, 36.79, calculate $\Delta G°$ and K for each of the following reactions at 25° C:

(a) $2Hg(l) + 2AgCl(s) \rightarrow 2Ag(s) + Hg_2Cl_2(s)$

(b) $2Hg(l) + 2Ag^+ \rightarrow 2Ag(s) + Hg_2^{++}$

EIGHT

Cation Group 2. The Acid Hydrogen Sulfide Group: Hg^{++}, Pb^{++}, Bi^{3+}, Cu^{++}, Cd^{++}, As(III), As(V), Sb(III), Sb(V), Sn(IV), Sn^{++}

8.1   Introduction

The positions of the members of this analytical group and of less common elements that also form sulfides from solutions that are 0.1–0.3 M in hydrochloric acid are shown below. Silver sulfide would also precipitate if silver ion had not previously been removed by chloride in Cation Group 1.

| | | VIII | | | | | IIIA | IVA | VA | VIA |
VIB	VIIB				IB	IIB				
					Cu^{II}			Ge^{II}	As^{III} As^{V}	Se^{IV*}
Mo^{VI}	Tc	Ru^{III}	Rh^{III}	Pd^{II}		Cd^{II}		Sn^{IV} Sn^{II}	Sb^{III} Sb^{V}	Te^{IV*}
	Re^{VII}	Os^{IV}	Ir^{III}	Pt^{IV}		Hg^{II}		Pb^{II}	Bi^{III}	Po^{II}

*Se and Te are precipitated as the free elements, not the sulfides, by H_2S.

143

8.2 Some Features of the Group as a Whole

The most notable common characteristic of all these elements is their affinity for sulfur. This is shown not only by their precipitation together as a group but also by the occurrence in nature of many sulfide ores, such as PbS, galena; $CuFeS_2$, chalcopyrite; HgS, cinnabar; and Sb_2S_3, stibnite.

None of the simple positive ions has noble gas structure. The ions Hg^{++} and Cd^{++} have 10 more outer-shell electrons than the noble gas octet, a d^{10} structure; Cu^{++} has the d^9 structure; and the other simple ions, Pb^{++}, Sn^{++}, Bi^{3+}, have 10 d electrons next to an outermost shell of a pair of s electrons, the $d^{10}s^2$ configuration. Such ions tend to form partially covalent bonds with I^-, O^{--}, and especially with S^{--} ions. Their compounds with these anions are frequently colored or black (Sec. 7.1); for example, yellow CdS and As_2S_3; red AsI_3, Pb_3O_4, and HgS; black BiI_3, SnO, and PbS. In combination with less polarizable anions, such as Cl^- or NO_3^-, the d^{10} and $d^{10}s^2$ cations are too stable to be excited by visible light, so they give colorless compounds, whereas Cu^{++}, with an incomplete d subshell, gives a colored chloride, and its hydrated salts are blue.

Arsenic, antimony, and tin are on the borderline between metals and non-metals in the periodic table. Their chlorides and hydrides are molecular substances. Their sulfides are acidic and dissolve in strong bases. Instead of forming positive ions, especially in their highest oxidation state, they form complexes with oxide, hydroxide, sulfide, and halide ions.

8.3 Precipitation of the Group

An extended discussion of the choice of conditions for precipitation is given in Secs. 2.4–2.6, and only a brief summary need be given here. The sulfides of this group have lower solubility products than those of later groups and require only a very low sulfide ion concentration for their precipitation. In a saturated solution of hydrogen sulfide, whether prepared from the gas or by hydrolysis of thioacetamide, the sulfide ion concentration varies inversely as the square of the hydrogen ion concentration:

$$K_{ip} = [H^+]^2[S^{--}] \tag{2.2}$$

When the hydrogen ion concentration is kept between 0.1 and 0.3 M, the sulfide ion concentration is at a sufficiently low level to precipitate cadmium sulfide, the most soluble of Cation Group 2,

$$[Cd^{++}][S^{--}] > K_{sp}$$

and not to precipitate zinc sulfide, the least soluble of Cation Group 3,

$$[Zn^{++}][S^{--}] < K_{sp}$$

During adjustment of acidity, copper(II) ion may be observed to form the deep blue complex with excess ammonia:

$$Cu^{++} + 4NH_3 \leftrightharpoons Cu(NH_3)_4{}^{++}$$

and the oxychlorides of antimony(III) and bismuth may precipitate; for example,

$$Bi^{3+} + Cl^- + H_2O \leftrightharpoons BiOCl(s) + 2H^+$$

The following equation is typical of the reaction of the simple ions with hydrogen sulfide:

$$Pb^{++} + H_2S \rightarrow PbS(s) + 2H^+$$

Molecular substances, solids, or complexes are also converted to the sulfides; for example,

$$HgCl_2 + H_2S \rightarrow HgS(s) + 2H^+ + 2Cl^-$$

$$2H_3AsO_3 + 3H_2S \rightarrow As_2S_3(s) + 6H_2O$$

$$2SbOCl(s) + 3H_2S \rightarrow Sb_2S_3(s) + 2H_2O + 2H^+ + 2Cl^-$$

$$SnCl_6{}^{--} + 2H_2S \rightarrow SnS_2(s) + 4H^+ + 6Cl^-$$

Arsenic(V) in H_3AsO_4 is reduced by thioacetamide to As_2S_3 and elemental S.

8.4   Subdivision of the Group

Because this is a very large group, it is convenient to subdivide it on the basis of the basic or acidic character of the sulfides. A strong base, such as 0.5 M KOH, dissolves the acidic sulfides of arsenic, antimony, and tin with the formation of oxy and thio anions; for example,

$$As_2S_3(s) + 6OH^- \rightarrow AsS_3{}^{3-} + AsO_3{}^{3-} + 3H_2O$$

$$SnS_2(s) + OH^- \rightarrow SnS_2OH^-$$

Tin should be in its 4+ oxidation state, for SnS is less acidic than SnS_2 and less soluble in KOH. The only other sulfide with acidic character is that of mercury(II), but it is so weak an acid that its solubility is small as long as the concentration of base is kept low.

8.5   Separation and Identification of Mercury

The analysis of the Copper Section starts with the treatment of the sulfides insoluble in KOH—namely, HgS, PbS, Bi_2S_3, CuS, and CdS—with 3 M HNO_3. All dissolve except the very insoluble sulfide of mercury(II) ($K_{sp} = 3 \times 10^{-52}$). Balanced equations for these reactions can be derived by the ion–electron method (Sec. 1.12) from the knowledge that the sulfide is oxidized to the metal ion and sulfur and nitrate ion is reduced to nitric oxide; that is, we start with $CuS(s) \rightarrow Cu^{++} + S$ and $NO_3^- \rightarrow NO(g)$, balance these partial equations, and then combine to get the complete equation. The oxidation of sulfide to sulfur reduces the sulfide ion concentration to such an extent that an unsaturated solution of the sulfide is produced

$$[Cu^{++}][S^{--}] < K_{sp}$$

and the precipitate dissolves.

Aqua regia must be used to dissolve HgS, for this reagent not only oxidizes sulfide but also removes Hg^{++} to give weakly dissociated $HgCl_4^{--}$, thereby decreasing the product of concentrations below the equilibrium value. The chemical equation can be obtained by balancing and combining the partials:

$$HgS(s) + Cl^- \rightarrow HgCl_4^{--} + S(s)$$

and

$$NO_3^- \rightarrow NO_2$$

Mercury is identified in the solution by reduction with tin(II) chloride to white Hg_2Cl_2 or a grey mixture of Hg_2Cl_2 and Hg. Nitrate and excess chloride interfere and have to be removed by evaporation prior to the test. The equations have been given in Sec. 7.3.

8.6 Separation and Identification of Lead

The nitric acid solution of the sulfides contains lead and nitrate ions from which the lead can be separated as lead sulfate. This precipitate is appreciably soluble in nitric acid because of the reactions

$$Pb^{++} + NO_3^- \leftrightharpoons PbNO_3^+ \qquad K = 14\ M^{-1}$$
$$H^+ + SO_4^{--} \leftrightharpoons HSO_4^- \qquad K = 98\ M^{-1}$$

In concentrated sulfuric acid solutions, nitric acid exists largely in the molecular form and can be driven off by evaporation. After the nitric acid and most of the water are gone, the temperature rises on further heating to 340°, and then sulfuric acid decomposes, giving off dense white fumes of sulfur trioxide:

$$H_2SO_4 + NO_3^- \xrightarrow{\Delta} HNO_3(g) + HSO_4^-$$
$$H_2SO_4 \xrightarrow{\Delta} SO_3(g) + H_2O(g)$$

The appearance of these fumes is thus a signal that nitric acid has been removed.

The lead sulfate that forms during this evaporation is often contaminated with small amounts of bismuth and copper sulfates. Lead sulfate, unlike these impurities, dissolves in hot $3\ M$ ammonium acetate:

$$PbSO_4(s) + 3OAc^- \rightarrow Pb(OAc)_3^- + SO_4^{--}$$

Lead is finally identified by precipitation as the chromate:

$$Pb(OAc)_3^- + CrO_4^{--} \rightarrow PbCrO_4(s) + 3OAc^-$$

8.7 Separation and Identification of Bismuth

Bismuth, after the removal of lead, is separated from copper and cadmium by addition of an excess of ammonia. Bismuth is precipitated as the hydroxide

$$Bi^{3+} + 3NH_3 + 3H_2O \rightarrow Bi(OH)_3(s) + 3NH_4^+$$

Although copper and cadmium hydroxides are insoluble in water, they do not form, because the concentrations of the cations are reduced by formation of weakly dissociated ammonia complexes:

$$Cu^{++} + 4NH_3 \leftrightharpoons Cu(NH_3)_4^{++} \qquad K = 1.2 \times 10^{12} \ M^{-4}$$
$$Cd^{++} + 4NH_3 \leftrightharpoons Cd(NH_3)_4^{++} \qquad K = 5.2 \times 10^6 \ M^{-4}$$

The conditions for unsaturated solutions are thereby established:

$$[Cu^{++}][OH^-]^2 < K_{sp}$$

and

$$[Cd^{++}][OH^-]^2 < K_{sp}$$

The identification of bismuth is based on reduction of the hydroxide to black bismuth metal by stannite. The principal reactions and products are expressed in the skeleton:

$$Bi(OH)_3(s) + Sn(OH)_3^- \rightarrow Sn(OH)_6^{--} + Bi(s)$$

Note that it is not balanced—why? Separate into partial equations and balance by the ion–electron method; note that the reaction is carried out in basic solution.

8.8 Identification of Copper and Cadmium

If the solution from the separation of bismuth hydroxide is blue due to $Cu(NH_3)_4^{++}$, this is usually sufficient indication of the presence of copper(II). When this test is indecisive, the ammonia complex is converted to copper(II) ion by acetic acid, and reddish copper(II) ferrocyanide is precipitated. You should be able to complete and balance the following:

$$Cu(NH_3)_4^{++} + HOAc \rightarrow ?$$

and

$$Cu^{++} + Fe(CN)_6^{4-} \rightarrow ?$$

Cadmium ferrocyanide could also form, but it is white.

The yellow color of CdS is characteristic enough to use for identification of cadmium, but it is easily masked by black CuS. To separate copper from cadmium prior to precipitation of the sulfide, we use selective reduction of copper by dithionite (also called "hyposulfite" or "hydrosulfite"). The principles of this test are discussed in Sec. 2.12. In the absence of Cu^{++} ion, the precipitation of yellow cadmium sulfide can be observed:

$$Cd(NH_3)_4^{++} + H_2S \rightarrow CdS(s) + 2NH_3 + 2NH_4^+$$

8.9 Subdivision of the Sulfides of the Arsenic Section

Analysis of the Arsenic Section starts with recovery of the sulfides from the alkaline solution containing the thio and oxy anions of arsenic, antimony, and tin(IV). Small amounts of mercury(II) may also be present.[1] The sulfides are precipitated by acidifying the solution, because this reduces the hydroxide ion concentration and reverses the reaction that produced the anions; for example,

$$SnS_2OH^- + H^+ \rightarrow SnS_2(s) + H_2O$$

$$SbS_3^{3-} + SbO_3^{3-} + 6H^+ \rightarrow Sb_2S_3(s) + 3H_2O$$

The sulfides of antimony and tin are dissolved by 6–8 M HCl:

$$Sb_2S_3(s) + 6H^+ + 8Cl^- \rightarrow 2SbCl_4^- + 3H_2S(g)$$

$$SnS_2(s) + 4H^+ + 6Cl^- \rightarrow SnCl_6^{--} + 2H_2S(g)$$

Arsenic(III) sulfide is virtually insoluble in the dilute acid.

8.10 Identification of Arsenic

Arsenic sulfide is dissolved by oxidizing sulfide to sulfur and arsenic(III) to arsenic(V) using hydrogen peroxide in ammoniacal solution. The equation can be derived by the ion–electron method from the following skeleton:

$$HO_2^- + As_2S_3(s) \rightarrow S(s) + OH^- + AsO_4^{3-}$$

The arsenate is then precipitated as magnesium ammonium arsenate

$$6H_2O + Mg^{++} + NH_4^+ + AsO_4^{3-} \rightarrow MgNH_4AsO_4 \cdot 6H_2O(s)$$

"Magnesia mixture," the reagent used for this purpose, is an ammoniacal solution of magnesium nitrate that contains sufficient ammonium nitrate to prevent the precipitation of magnesium hydroxide (Sec. 2.7). It supplies high

---

[1] Mercury will appear in the Arsenic Section only if the other members of this section are present. In small amounts, reprecipitated HgS dissolves in 12 M HCl; any that remains with arsenic sulfide is unaffected by NH_3 and H_2O_2. Mercury(II) is reduced to the free metal by aluminum and does not interfere with the tin test. Its sulfide will precipitate with that of antimony but can be distinguished from it by its insolubility in 6 M HCl or ammonium sulfide.

concentrations of magnesium and ammonium ions and is sufficiently basic to keep a substantial portion of the arsenate as AsO_4^{3-} rather than $HAsO_4^{--}$. When these requirements are fulfilled, the condition for precipitation is established:

$$[Mg^{++}][NH_4^+][AsO_4^{3-}] > K_{sp}$$

The precipitate is dissolved in acetic acid, which removes arsenate ion to form weakly ionized $HAsO_4^{--}$ or $H_2AsO_4^-$:

$$MgNH_4AsO_4 \cdot 6H_2O(s) + 2HOAc \rightarrow$$
$$Mg^{++} + NH_4^+ + H_2AsO_4^- + 2OAc^- + 6H_2O$$

Arsenic is finally identified by precipitation of pale reddish brown silver arsenate:

$$3Ag^+ + H_2AsO_4^- \rightarrow Ag_3AsO_4(s) + 2H^+$$

8.11   Identification of Antimony

Part of the hydrochloric acid solution containing tin and antimony is treated with oxalic acid and thioacetamide. Both tin(IV) and antimony(III) form oxalate complexes:

$$SbCl_4^- + 3H_2C_2O_4 \rightarrow Sb(C_2O_4)_3^{3-} + 4Cl^- + 6H^+$$
$$SnCl_6^{--} + 3H_2C_2O_4 \rightarrow Sn(C_2O_4)_3^{--} + 6Cl^- + 6H^+$$

Dissociation of the tin complex is so limited that SnS_2 cannot be precipitated from this solution, but precipitation of Sb_2S_3 is not inhibited:

$$2Sb(C_2O_4)_3^{3-} + 3H_2S \rightarrow Sb_2S_3(s) + 6HC_2O_4^-$$

The deep orange color of this sulfide is distinctive enough to identify the presence of antimony.

8.12   Identification of Tin

Tin is identified by the reducing action of Sn^{++} ion on $HgCl_2$, the same reaction used to identify mercury (Secs. 7.3 and 8.5). Since the reducing agent is in limited supply, reduction of mercury(II) chloride goes only to white mercury(I) chloride. The hydrochloric acid solution of tin and antimony contains tin(IV) as a chloro complex, which must be reduced to tin(II) before the test is made. Aluminum will do this:

$$3SnCl_6^{--} + 2Al(s) \rightarrow 2Al^{3+} + 3Sn^{++} + 18Cl^-$$

but it will also reduce antimony and some tin to the free elements. You can derive balanced equations from the following skeletons:

$$Sn^{++} + Al(s) \rightarrow Sn(s) + Al^{3+}$$
$$SbCl_4^- + Al(s) \rightarrow Sb(s) + Al^{3+}$$

Tin, but not antimony, redissolves in hot hydrochloric acid after the aluminum is all consumed:

$$Sn(s) + 2H^+ \rightarrow H_2(g) + Sn^{++}$$

8.13 Illustrative Numerical Calculations

A problem dealing with precipitation of sulfides was worked through in Sec. 2.6, and the separation of copper from cadmium was considered in Sec. 2.12.

(a) Fractional Precipitation of Sulfides

Suppose we have a solution that is $0.010\ M$ in both lead(II) and iron(II) ions, and we wish to separate these ions by fractional precipitation of the sulfides. At $100°C$, sulfide is gradually generated in solution by hydrolysis of thioacetamide.

A much lower concentration of sulfide is required to precipitate PbS than to precipitate FeS:

$$PbS: \quad [S^{--}] = \frac{K_{sp}(PbS)}{[Pb^{++}]} = \frac{2.6 \times 10^{-23}}{1.0 \times 10^{-2}} = 2.6 \times 10^{-21}\ M$$

$$FeS: \quad [S^{--}] = \frac{K_{sp}(FeS)}{[Fe^{++}]} = \frac{1.3 \times 10^{-16}}{1.0 \times 10^{-2}} = 1.3 \times 10^{-14}\ M$$

Lead(II) sulfide will therefore precipitate first and continue to do so until the sulfide ion concentration has increased beyond $1.3 \times 10^{-14}\ M$. At this point the lead remaining in solution is at a concentration given by

$$[Pb^{++}] = \frac{2.6 \times 10^{-23}}{1.3 \times 10^{-14}} = 2.0 \times 10^{-9}\ M$$

so that only $(2.0 \times 10^{-9} \times 100)/(1.0 \times 10^{-2})$ or $2.0 \times 10^{-5}\%$ is unprecipitated.

We can prevent precipitation of FeS altogether by adjusting the hydrogen ion concentration so that $[S^{--}]$ cannot exceed $1.3 \times 10^{-14}\ M$. The concentration of hydrogen ion can then be calculated from the ion product (Sec. 2.6):

$$[H^+]^2[S^{--}] = 1.1 \times 10^{-18}\ M^3 \quad \text{or} \quad [H^+] = \left(\frac{1.1 \times 10^{-18}}{1.3 \times 10^{-14}}\right)^{1/2}$$

whence $[H^+] = 9 \times 10^{-3}\ M$. Clearly, the concentration of hydrochloric acid used in precipitation of Group 2, namely, 0.1–$0.3\ M$, will ensure that FeS does not precipitate.

(b) Subdivision of the Group

The equilibrium constant for the reaction of SnS_2 with base

$$SnS_2(s) + OH^- \leftrightharpoons SnS_2OH^- \qquad K = \frac{[SnS_2OH^-]}{[OH^-]}$$

is reported to be 1.6. Let S be the molar solubility in 0.30 M KOH (allowing for some dilution of the 0.5 M reagent). From the stoichiometric relations

$$S = [SnS_2OH^-] \quad \text{and} \quad 0.30 - S = [OH^-]$$

we have $1.6 = S/(0.30 - S)$, whence $S = 0.18$ M.

(c) Dissolution of Sulfides in Nitric Acid

Equilibrium constants for reactions of the type

$$3MS(s) + 8H^+ + 2NO_3^- \leftrightharpoons 2NO(g) + 4H_2O(l) + 3M^{++} + 3S$$

cannot be obtained by direct analysis because of the difficulty of preparing suitable equilibrium mixtures. We can calculate the constants by making use of their thermodynamic relationship with $E°$ or $\Delta G°$; for example,

$$\Delta G° = -2.303 RT \log K \tag{1.3}$$

where the factor $2.303 RT$ is 1.364 kcal mol^{-1} at 25°C.

From tabular values of standard Gibbs energies of formation we can compute $\Delta G°$, bearing in mind the convention that $\Delta G_f°$ is zero for H^+ and rhombic sulfur:

$$\begin{aligned}
\Delta G° &= 2\Delta G_f°(NO) + 4\Delta G_f°(H_2O) + 3\Delta G_f°(M^{++}) - 3\Delta G_f°(MS) \\
&\quad - 2\Delta G_f°(NO_3^-) \\
&= 2(20.72) + 4(-56.69) + 3[\Delta G_f°(M^{++}) - \Delta G_f°(MS)] - 2(-49.37) \\
&= -86.58 \text{ kcal} + 3[\Delta G_f°(M^{++}) - \Delta G_f°(MS)]
\end{aligned}$$

For HgS, $\Delta G_f°(M^{++}) - \Delta G_f°(MS) = 39.38 - (-10.22)$, making $\Delta G°$ for dissolution of black HgS in dilute nitric acid equal to $+62.22$ kcal. For PbS the difference is $-5.81 - (-22.15)$, making $\Delta G°$ for its dissolution -37.56 kcal. A positive value for $\Delta G°$ means that at constant temperature and pressure the dissolution reaction is uphill or not spontaneous. Our computation, in other words, indicates that HgS will not dissolve in nitric acid whereas PbS will, and this checks with our laboratory experience.

The equilibrium constants tell us much the same thing:

$$\text{HgS:} \quad \log K = \frac{-62.22}{1.364} \doteq -45.6 \quad \text{or} \quad K = 2.5 \times 10^{-46}$$

$$\text{PbS:} \quad \log K = \frac{-(-37.56)}{1.364} = 27.6 \quad \text{or} \quad K = 4 \times 10^{27}$$

Commentary   The direction of change can sometimes be altered by changes in concentration. Here the values of the constants are so small and so large that moderate changes in concentration of nitric acid are unlikely to have any effect on the separation. Thermodynamic calculations give us no inkling of how fast the reaction will go. We have to warm the mixture to make it go fast enough to be useful.

(d) Separation by Complex Formation

After separation of lead sulfate in the Copper Section, the solution may contain about one drop of 18 M H_2SO_4 in excess. This is neutralized by 15 M NH_3, and about one drop of the base is added in excess. With dilution and washings taken into account, the total volume of solution at this point is perhaps 10 drops, or 0.50 ml. In the practice solution we start with a sample containing 2 mg of Cd^{++} ion. Some is lost at each stage because our separations are not quantitative; suppose half, or 1 mg, remains. We can then compute the following concentrations:

$$[NH_4^+] = 3.6\ M; \quad [NH_3] = 1.5\ M; \quad \text{total Cd} = 1.8 \times 10^{-2}\ M$$

For simplicity we suppose that virtually all the cadmium goes into the tetrammine complex:

$$Cd^{++} + 4NH_3 \leftrightharpoons Cd(NH_3)_4^{++} \qquad \beta_4 = \frac{[Cd(NH_3)_4^{++}]}{[Cd^{++}][NH_3]^4} = 1.9 \times 10^7\ M^{-4}$$

Using $[Cd(NH_3)_4^{++}] = 1.8 \times 10^{-2}\ M$ and $[NH_3] = 1.5\ M$, we can solve this for the concentration of residual cadmium ions:

$$[Cd^{++}] = \frac{[Cd(NH_3)_4^{++}]}{\beta_4[NH_3]^4} = \frac{1.8 \times 10^{-2}}{1.9 \times 10^7 \times (1.5)^4} = 1.9 \times 10^{-10}\ M$$

The hydroxide ion concentration of the mixture is controlled by K_b of ammonia:

$$[OH^-] = K_b \times \frac{[NH_3]}{[NH_4^+]} = 1.8 \times 10^{-5} \times \frac{1.5}{3.6} = 7.5 \times 10^{-6}\ M$$

The product of concentrations of cadmium and hydroxide ions is less than the solubility product of $Cd(OH)_2$:

$$1.9 \times 10^{-10} \times (7.5 \times 10^{-6})^2 < 4 \times 10^{-15}\ M^3$$

so that cadmium hydroxide cannot precipitate. Cadmium sulfide, on the other hand, is much less soluble; to get a precipitate of it the sulfide ion concentration must be greater than

$$[S^{--}] = \frac{7 \times 10^{-27}}{1.9 \times 10^{-10}} = 3.7 \times 10^{-17}\ M$$

It is easy to get concentrations higher than this by adding H_2S or thioacetamide to the basic solution. Precipitation of yellow CdS is the final identification test for cadmium.

Commentary The solution we are dealing with is fairly concentrated, whereas values of most equilibrium constants are strictly applicable to dilute solutions. The value of β_4 used here is a special one appropriate for 2 M NH_4NO_3 solutions; for very dilute solutions, β_4 is slightly smaller. We do not know the value of this constant in the mixture under examination, but it is unlikely that it is in error enough to affect our predictions.

EXPERIMENTAL PART

8.14   Analysis of Known and Unknown Samples

(*a*) Use 5 drops of a *known* or *practice* solution that contains 2 mg each of Hg^{++}, Bi^{3+}, Cu^{++}, As(III), and Sb(III), 4 mg of Pb^{++}, and 8 mg each of Cd^{++} and Sn(IV) per ml. If it contains a sediment, shake well and quickly withdraw the sample.

(*b*) The unknown may be (1) a solution containing the ions of this and possibly other groups or (2) a solid mixture. If it is Solution 1.1 from Outline 1, use all of it. If it is a special unknown solution on this group alone, treat it like the practice solution. If it is a solid, dissolve about 20 mg of a representative sample (T-1) in 10 drops of 6 *M* HCl and use half for the analysis.

8.15   Preliminary Preparations

P-1.   Calibration of a Test Tube

Measure 2.5 ml of water into a dry test tube. Mark the position of the bottom of the meniscus with a strip of label.

P-2.   Oxidation of Tin(II)

Transfer the solution to be analyzed, which should be acid, to a small casserole, add 2 drops of 3 % H_2O_2, and boil down the solution to a volume of a few drops. Do this by using a microflame [T-7(a, b)], swirling the solution constantly, and withdrawing the dish from the flame somewhat before the final volume is reached —the solution will continue to evaporate as it cools. The chlorides of As, Sb, Sn, and Hg are sufficiently volatile that some of these elements may be lost if evaporation is carried too far.

Dilute the remaining solution with 3–5 drops of water, and transfer it to the calibrated test tube. Rinse out the casserole walls and bottom twice with 3–5 drops of water, and transfer both rinses to the test tube. Do not be concerned if the solution is turbid. This is usually caused by SbOCl or BiOCl, which are readily converted to the sulfides in a later step.

P-3.   Adjustment of Acidity

Add 6 *M* NH_3 to the solution until it is basic to litmus or wide-range indicator paper. If Cu^{++} ion is present, the formation of the deep blue ammonia complex also acts as an indicator, but the blue color may be obscured by colored precipitates or complex ions of Group 3. To the basic solution, add 2 *M* HCl drop by drop until the pH is between 2 and 3 according to wide-range indicator paper. If the solution is made too acid, bring it back with a fraction of a drop of 6 *M* NH_3 [T-4(b)]. Then add exactly 0.35 ml of 2 *M* HCl by means of your calibrated pipet. Add 4 drops of 13 % thioacetamide solution (hereafter abbreviated TA) and enough water to make a total volume of 2.5 ml. Mix well [T-5(b, c)]. Check the acidity against short-range indicator paper. It should have a pH of 0.5 corresponding to a hydrogen ion concentration of 0.3 *M*. Treat the solution further by Outline 2.

OUTLINE 2. Precipitation and Subdivision of Cation Group 2

Precipitation of the group: Warm the solution contg TA at pH 0.5 from P-2 in a bath of vigorously boiling water for 5 minutes.[1] Centrifuge and withdraw soln to another test tube. Check the acidity with short-range indicator paper.[2] Add 2 drops TA and heat several minutes in the boiling water. Centrifuge if more ppt appears. Continue heating with TA as long as ppt is obtained. Then dilute several drops of the clear soln with twice its vol of water, add 1 drop TA, and heat again. If a ppt appears, treat the whole soln this way; divide it between several test tubes if necessary.[3]

Combine ppts if several portions have been obtained (T-11). Wash the pptd sulfides twice with about 20 drops hot water[4] contg 1 drop 6 M NH$_4$Cl (T-10).[5] Add the first washing to Soln 2.1 in a general analysis for all groups.

Precipitate 2.1. HgS, PbS, Bi$_2$S$_3$, CuS, CdS, As$_2$S$_3$, Sb$_2$S$_3$, SnS$_2$, S. Add 10 drops 0.5 M KOH and heat briefly in the water bath. Centrifuge and transfer soln to a centrifuge tube. Repeat the extn with a second portion of KOH soln. Combine solns.

Solution 2.1. Ions of later groups, HCl, NH$_4$Cl, H$_2$S, TA. Discard if original sample contained only ions of Group 2. If soln is to be analyzed for ions of later groups, add 1 ml 12 M HCl. Transfer to a casserole and evap almost to dryness.[6] Add several drops water and 1 drop 6 M HCl and reserve in a stoppered tube for Group 3.

Precipitate 2.2. HgS, PbS, Bi$_2$S$_3$, CuS, CdS, S. Wash twice with hot water to which 1 drop 0.2 M NH$_4$NO$_3$ has been added.[7]

Analyze according to **Outline 3.**[8]

Solution 2.2. AsS$_3^{3-}$, AsO$_3^{3-}$, SbS$_3^{3-}$, SbO$_3^{3-}$, SnS$_3^{--}$, SnS$_2$OH$^-$, (HgS$_2^{--}$), KOH. Recentrifuge to remove last traces of Ppt 2.2.[9] Transfer the clear, yellow soln to a test tube.

Analyze according to **Outline 4.**[10]

Notes on Outline 2

1. Allow sufficient time for the hydrolysis of TA. Heating also helps to coagulate the sulfides so that they are easier to settle by centrifugation.

2. Hydrogen ions are liberated in the precipitation of the sulfides; for example, $Cu^{++} + H_2S \rightarrow CuS(s) + 2H^+$. If the solution becomes too acidic, ammonia must be added to bring the pH back to 0.5.

3. The diluted solution will have a hydrogen ion concentration of 0.1 (pH 1). The dilution may be necessary to obtain complete precipitation of CdS. If a bright yellow precipitate forms that is soluble in 6 M HCl but insoluble in KOH, it is CdS, and no further test for Cd^{++} need be made. If the concentration of chloride ion is unusually high, black PbS may also precipitate.

4. Use water from a test tube in the hot water bath or add cold water and heat in the water bath for a few minutes. Washing is necessary to remove excess H_2S, which increases the solubility of HgS in KOH, together with ions of later groups, if present.

5. Ammonium chloride prevents the sulfides from turning to a colloidal suspension.

6. This evaporation is necessary to remove H_2S and TA, which are slowly oxidized by atmospheric oxygen to sulfur and even sulfate. The latter would precipitate Ba^{++} and Sr^{++} ions and cause them to be lost.

7. If a colloidal suspension persists despite the presence of NH_4NO_3, add a few crystals of the solid salt. Sometimes a repetition of this treatment is necessary, but keep the addition of NH_4NO_3 to the minimum required to flocculate the colloid.

8. If analysis of the Copper Section cannot be done at once, cover the sulfides with water and a drop of TA, and stopper the tube with a cork or medicine dropper bulb. It is better, if possible, to carry the analysis as far as the separation of Precipitate 3.1 and Solution 3.1 in Outline 3.

9. Complete separation of solution and precipitate is difficult in a test tube. Hence a second centrifugation and separation are made in a centrifuge tube. If a colloid persists, add a few crystals of NH_4NO_3 or a drop of 0.2 M solution and warm to coagulate it.

10. If this solution cannot be analyzed at once, add a drop of TA and preserve it in a tightly stoppered tube. A precipitate sometimes appears after the solution has stood for a day or so. If it is colored, it is a sulfide. Its formation does no harm; keep it with the solution.

OUTLINE 3. The Systematic Analysis of the Copper Section
of Cation Group 2

Precipitate 2.2, Outline 2. HgS, PbS, Bi$_2$S$_3$, CuS, CdS, S. Treat ppt with 5 drops water and an equal vol 6 M HNO$_3$. Warm in water bath until reaction occurs.[1] Centrifuge and wash ppt once with water. Add washings to Soln 3.1.

Precipitate 3.1. HgS (black), S colored by traces of sulfides, or 2HgS·Hg(NO$_3$)$_2$ (white).[2] Add a few drops aqua regia and warm. Transfer soln to crucible.[3] Evap over microflame to a small vol but not to dryness.[4] Dilute with about 1 ml water[5] and transfer to centrifuge tube. Centrifuge if not clear and discard residue. Add SnCl$_2$ soln.

White ppt, Hg$_2$Cl$_2$, turning grey, Hg + Hg$_2$Cl$_2$: **presence of Hg(II).**

Solution 3.1. Pb^{++}, Bi^{3+}, Cu^{++}, Cd^{++}, H$^+$, NO$_3$$^-$. Transfer to a casserole and add 2 drops 18 M H$_2$SO$_4$. Evap over a microflame and directly under a ventilating duct or in the hood until *dense, choking*, white fumes of SO$_3$ are copiously evolved.[6] Cool for several minutes. Cautiously add 5 drops water; rub walls and bottom of the casserole with a rod to loosen caked ppt. Transfer soln quickly to a centrifuge tube. Rinse casserole with two small portions of water and add these to centrifuge tube. Centrifuge and withdraw soln to another centrifuge tube. Wash ppt with water.

Precipitate 3.2. PbSO$_4$ [(BiO)$_2$SO$_4$].[7] Warm with a few drops NH$_4$OAc soln. Centrifuge and discard residue. To soln add 1 drop 6 M HOAc and a few drops of K$_2$CrO$_4$ soln.

Strong yellow ppt, PbCrO$_4$: **presence of Pb.**[8]

Solution 3.2. Bi^{3+}, Cu^{++}, Cd^{++}. Add 15 M NH$_3$ until soln is alkaline. Check with litmus. Centrifuge even though you see no ppt.[10]

Blue soln indicates **presence of Cu.**

Precipitate 3.3. Bi(OH)$_3$. Add a few drops of freshly prepared sodium stannite reagent.[11]

Immediate change to *jet black*, Bi: **presence of Bi.**

Solution 3.3. Cu(NH$_3$)$_4$$^{++}$, Cd(NH$_3$)$_4$$^{++}$. If soln is not blue, add TA and warm. Yellow ppt, CdS: **presence of Cd.**

If soln is blue, add one-half spatula-full of Na$_2$S$_2$O$_4$.[12] Warm for 1–2 minutes.[13] Centrifuge quickly and remove soln.[14] Repeat the treatment with Na$_2$S$_2$O$_4$ as long as Cu forms.[15] To the Cu-free soln add TA and warm.

Yellow ppt, CdS: **presence of Cd.**[16,17]

Notes on Outline 3

1. There is an induction period during which no apparent change occurs. Then bubbling starts and the bubbles often carry sulfur and sulfides to the surface. Absence of bubbling does not indicate no reaction.

2. As these formulas and colors imply, a black precipitate at this stage is not necessarily an indication of the presence of mercury, nor is the formation of a white or yellow precipitate proof of its absence. The precipitate must always be dissolved and the mercury must be sought in the solution.

3. In drawing off the solution, leave behind any yellow particles of sulfur.

4. Mercury(II) chloride is somewhat volatile and should not be heated too strongly.

5. Evaporation destroys nitrate and removes excess HCl. The concentration of acid is further decreased by dilution, because the reaction between $HgCl_2$ and Sn^{++} ion is slow if the concentration of HCl is too high.

6. The formation of SO_3 is a signal that nitrate has been removed as molecular HNO_3. Once seen, fumes of SO_3 are easy to recognize thereafter. They are heavier and more irritating than the fumes of nitric acid that come off first. Failure to evaporate until copious fumes are evolved may cause enough lead to stay in solution to spoil subsequent tests. Yet evaporation should not be carried to dryness, because the anhydrous sulfates come back into solution slowly. If this should happen, add 2 drops of 6 M H_2SO_4 and 3 drops of water, and warm the casserole on the water bath for several minutes. Stir to disperse the residue in the solution.

7. Formation of a precipitate at this point is not a certain indication of the presence of Pb. Occasionally other sulfates, such as $(BiO)_2SO_4$, precipitate here. The bismuth compound is less soluble than $PbSO_4$ in NH_4OAc, and $(BiO)_2CrO_4$ is more soluble than $PbCrO_4$ in HOAc and is insoluble in NaOH.

8. If a large quantity of bismuth is found, check the solubility of this precipitate in NaOH.

9. If the blue color is faint and further confirmation is required, acidify a drop of the solution with HOAc and add a drop of $K_4Fe(CN)_6$ solution. Copper(II) forms a red ferrocyanide, whereas that of cadmium is white.

10. Beginners find small, gelatinous precipitates of $Bi(OH)_3$ hard to see, particularly when suspended in a blue solution.

11. To a drop of $SnCl_2$ reagent, add 6 M NaOH drop by drop with careful mixing until the precipitate of $Sn(OH)_2$ redissolves. This will take at least 2 drops of NaOH solution. If there is doubt as to the effectiveness of the $SnCl_2$ solution, which is gradually oxidized by air, prepare some stannite from it and test its action on $Bi(OH)_3$ made from the standard solution of Bi^{3+} ion.

12. Sodium dithionite is a powerful reducing agent and is unstable. Keep it dry and away from heat. Use a dry spatula in measuring the solid. One-half spatula-full, or about 50 mg, should reduce 12.5 mg of Cu^{++} ion. The blue color of the copper ammonia complex should permanently disappear upon addition of dithionite.

13. Precipitated copper is red. Black or brown residues frequently obtained here are colored by other metals. This does no harm; in fact, it is an advantage to remove traces of the other metals before the cadmium test.

14. If the solution stands in contact with copper too long, some copper can redissolve.

15. After prolonged treatment with dithionite, a yellow coloration may appear. This is cadmium sulfide. Add TA at once and warm. The sulfide apparently comes from decomposition of the dithionite after removal of copper.

16. Occasionally a buff, green, or black precipitate is obtained at this point. This is a result of careless separations in previous steps. If the precipitate is large, remove the solution and discard it. Dissolve the precipitate as far as possible in one drop of 6 M HCl. Centrifuge and transfer the clear solution to a test tube. Dilute with 1 ml of water, add TA, and warm to reprecipitate the sulfide.

17. Cadmium may also be identified during the precipitation of the group. See Note 3 to Outline 2.

OUTLINE 4. The Systematic Analysis of the Arsenic Section
of Cation Group 2

Solution 2.2, Outline 2. AsS$_3$$^{3-}$, AsO$_3$$^{3-}$, SbS$_3$$^{3-}$, SbO$_3$$^{3-}$, SnS$_3$$^{--}$, SnS$_2OH^-$, (HgS$_2$$^-$), KOH. Add 1 drop TA.[1] Then very carefully add 2 M HCl until soln is just acid to litmus.[2] Note the color and character of the ppt.[3] Transfer mixture to a centrifuge tube and centrifuge.[4] Draw off supernatant liquid as completely as possible and discard it.[5]

Precipitate 4.1. As$_2$S$_3$ (yellow), Sb$_2$S$_3$ (orange), SnS$_2$ (yellow), (HgS, black to red), S. Add 5–10 drops 12 M HCl. Stir to loosen the sulfide cake and warm in the water bath only until reaction occurs.[6] Centrifuge and transfer supernatant soln to casserole.[7] Wash ppt with a few drops 6 M HCl and add washings to casserole.

Precipitation 4.2. As$_2$S$_3$, (HgS), S. Add 2 drops 15 M NH$_3$ and 1 drop 3% H$_2$O$_2$. Warm briefly and centrifuge. Discard residue.[8] To soln add several drops of magnesia mixture. Rub the inner wall of the tube with a stirring rod to induce crystallization. Let stand 5 minutes.[9] Centrifuge and discard soln. Wash ppt once with water.[10]

Precipitate 4.3. White, cryst MgNH$_4$AsO$_4$·6H$_2$O. Add 1 drop 6 M HOAc and several drops AgNO$_3$ soln.[11]
Reddish brown ppt, Ag$_3$AsO$_4$: **presence of As.**

Solution 4.2. SnCl$_6$$^{--}$, SbCl$_4$$^-$, (HgCl$_4$$^{--}$), HCl, H$_2$S, TA. Evap soln over a microflame to half the original vol or not less than 4 drops.[12] Dilute with about 1 ml water and divide between two test tubes.

Sb Test. Add one-fourth to one-half spatula-full solid H$_2$C$_2$O$_4$ and a few drops TA. Warm.
Deep orange[16] ppt, Sb$_2$S$_3$: **presence of Sb.**

Sn Test. Add a piece of Al wire about 5 mm long and follow with 10 drops 6 M HCl. Heat in water bath until Al is consumed and for a minute longer.[13] Transfer to a centrifuge tube and centrifuge.

Solution 4.4. Sn^{++}, Al^{3+}, HCl. Dilute with an equal vol of water. Without delay[15] add 1–2 drops HgCl$_2$ soln. White to grey ppt, Hg$_2$Cl$_2$ and Hg: **presence of Sn.**

Precipitate 4.4. Black flecks of Sb.[14]

Notes on Outline 4

1. Thioacetamide is required to supply H_2S that is lost by volatilization.

2. If the solution is made too acidic, some SnS_2 may be lost. Hence 2 M rather than 6 M HCl is used. Mix thoroughly after each addition.

3. The sulfides of this section are all highly colored. A white, hazy turbidity that does not settle readily upon centrifugation is sulfur in colloidal form. When only this is obtained, the arsenic section is absent.

4. The sulfides are precipitated in a test tube to allow careful neutralization and thorough mixing. Since they do not pack down well, separation of precipitate and solution is improved if a centrifuge tube is used.

5. The extraction with 12 M HCl in the next step will fail if the acid becomes diluted with excess water in the precipitate.

6. Note any change in character of the precipitate during this treatment. A bright yellow residue is usually As_2S_3; a black precipitate may contain HgS. A colorless solution and a white turbidity of S indicate the absence of arsenic. Do not prolong the heating of the mixture, for some As_2S_3 may dissolve slowly.

7. Draw off the supernatant solution through a wisp of cotton wrapped around the tip of the pipet. If the pipet or casserole contains water, this may decrease the concentration of HCl to such an extent that Sb_2S_3 reprecipitates. This does no harm.

8. A black residue is probably HgS. It can be identified by the procedure given under Precipitate 3.1, Outline 3. Ammonia readily dissolves As_2S_3 but not HgS; the H_2O_2 oxidizes As(III) to As(V).

9. This precipitate sometimes supersaturates badly. Give it time to form before reporting a negative result.

10. A washing here removes traces of chloride that would precipitate with silver nitrate in the next step.

11. If a white precipitate of AgCl forms, add more $AgNO_3$.

12. If evaporation is carried too far, some $SnCl_4$ may be lost.

13. Aluminum can reduce Sn(IV) and Sn(II) to grey Sn. It will dissolve in HCl after the Al has been consumed; Sb will not dissolve.

14. If the regular Sb test was indecisive, a confirmatory test can be made on this residue. Wash it with water and dissolve in a drop of 6 M HNO_3. Dilute with water, add TA, and warm to precipitate Sb_2S_3.

15. Since Sn^{++} is oxidized by oxygen of air, the addition of $HgCl_2$ should not be delayed.

16. The precipitate is frequently orange at first and then darkens by postprecipitation of impurities. If it is yellow, it may be As_2S_3 because the treatment of Precipitate 4.1 with HCl was too prolonged. Test its solubility in 12 M HCl and in $NH_3 + H_2O_2$. If the precipitate in the Sb test is very dark, it may be contaminated with HgS. Remove the solution. Add 6 M HCl to the precipitate and warm for a minute or so. Centrifuge and discard the residue. Dilute the solution, add TA, and warm to reprecipitate Sb_2S_3.

EXERCISES

8.1. Name the elements of this analytical group and tell why they occur in the group.

8.2. For each element in Group 2 give its position in the periodic table and its principal oxidation states.

8.3. Identify by formula: thioacetamide, aqua regia, sodium dithionite, arsenious acid, tetramminecopper(II) ion, thioarsenite ion, sodium stannite, hexachlorostannate(IV) ion.

8.4. Give the colors of: $CdCl_2$, $CuSO_4$, $BiOCl$, As_2S_3, $Cu_2Fe(CN)_6$, SnS_2, $CuSO_4 \cdot 5H_2O$, Sb_2S_3, $Cd(NH_3)_4Cl_2$, $PbSO_4$, H_3AsO_4, $SnCl_4$, $MgNH_4AsO_4 \cdot 6H_2O$, PbS, Hg_2Cl_2, CdS, CuS.

8.5. Name the substances in Exercise 8.4.

8.6. Fill in the blanks in the following table.

		Predominant forms in		
Element and oxidation state	*0.3 M HCl*	*6 M HCl*	*6 M NH_3*	*6 M NaOH*
Hg(II)			$HgO(s)$	
Pb(II)		$PbCl_3^-$	$Pb(OH)_2(s)$	
Bi(III)	$BiOCl(s)$	$BiCl_4^-$		
Cu(II)		$CuCl_4^{--}$		
Cd(II)		$CdCl_4^{--}$		
As(III)	H_3AsO_3	$AsCl_4^-$	AsO_3^{3-}*	
Sb(III)		$SbCl_4^-$		
Sn(IV)	$SnCl_6^{--}$	$SnCl_6^{--}$	SnO_2	

* May also exist in hydrated form, e.g., $As(OH)_4^-$.

8.7. Which of the following have some acidic character? CdS, $As(OH)_3$, SnS_2, HgO, $Bi(OH)_3$, As_2S_3. Represent such behavior by equations.

8.8. What are the white precipitates that form when bismuth and antimony trichlorides are added to water? Represent the hydrolysis reactions by equations. Explain why the precipitates dissolve in hydrochloric acid.

8.9. Write an equation for the hydrolysis of thioacetamide in acid solution. What is the role of H^+ in this reaction? Why is the formation of H_2S not suppressed by an excess of HCl?

8.10. By reference to appropriate equilibrium constants, explain how it is possible to separate Cd^{++} from Zn^{++} ion with hydrogen sulfide in 0.1–0.3 M HCl.

8.11. Cite examples from the analytical procedure of separations based on:
   (a) Formation of an ammonia complex ion.
   (b) The acidic nature of some sulfides.
   (c) Selective reduction of one ion.
   (d) Differing solubility products of some sulfates.

8.12. Write equations for the successive reactions of each ion of this group from initial precipitation to identification.

8.13. If silver were not precipitated in Group 1 but came down as the sulfide in Group 2, where would it appear in the course of the analysis?

8.14. What errors in technique might be responsible for:
   (a) Failure to obtain CdS from Solution 3.3.
   (b) Formation of a yellow precipitate in the antimony test.
   (c) Formation of a brown precipitate in the cadmium test.
   (d) Complete dissolution of Precipitate 2.2 by 3 M nitric acid even though HgS was present.
   (e) Failure to obtain Bi from $Bi(OH)_3$.
   (f) Failure to obtain Precipitate 4.3 even though arsenate was present.

8.15. Give a reason for each:
   (a) The use of freshly prepared sodium stannite.
   (b) The use of ammonium salts in the wash water when washing sulfides.
   (c) Dilution of the solution after removal of the first group precipitate.
   (d) The quick separation of copper from the solution after the dithionite treatment.
   (e) The use of oxalic acid in the antimony test.
   (f) The use of acetic acid in the lead test.
   (g) The use of a dilute solution for the reaction between tin(II) and mercury(III) chlorides.
   (h) Why chloride solutions of the ions of this group are never evaporated to dryness.

8.16. If you were given solutions of the following samples, what is the briefest procedure that could be used to answer the question posed in each part?
   (a) Is the heavy stone you picked up in a dry Australian stream bed cassiterite (SnO_2), a valuable source of tin?
   (b) Is the black powder found in an archeological dig and presumed to have been used as eyeshadow, kohl (black Sb_2S_3) or galena (PbS)?
   (c) Is a white powder, tartar emetic ($KSbOC_4H_4O_6 \cdot \frac{1}{2}H_2O$) or the gastrointestinal astringent bismuth subcarbonate $[(BiO)_2CO_3 \cdot \frac{1}{2}H_2O)]$?
   (d) Is the red pigment in a mural painting realgar (As_4S_4), vermilion (HgS), or red lead (Pb_3O_4)?

8.17. Devise the briefest possible procedure for analyzing each of the following samples:
   (a) A mixture of $Bi(OH)_3$ and $Cu(OH)_2$.
   (b) A solution containing Hg^{++} and Cd^{++} ions.
   (c) A solution containing Ag^+ and Cu^{++} ions.

(d) A solution containing Sn(IV) and Cd^{++} ion.

(e) A mixture of SnS_2, HgS, and CuS.

(f) A solution containing Hg^{++}, Pb^{++}, and Cd^{++} ions.

8.18. A solid is suspected of being either type metal (Pb and Sb with smaller amounts of Sn, Cd, or Hg) or a fusible alloy, such as is used in fire sprinkler systems (Bi and Pb with smaller amounts of Sn and Cd). Given a solution of the sample, what would be the briefest procedure for determining the constituents? You can use the procedures of the systematic analysis, introducing short cuts, or invoke other properties of the ions.

8.19. Calculate the maximum concentration of hydrogen ion that can be present in a cadmium nitrate solution saturated with hydrogen sulfide at 25°C and 1 atm pressure if no more than 0.0010 mg Cd^{++} per ml is to be left in solution. Take the solubility product of active cadmium sulfide to be 1.0×10^{-26} M^2.

8.20. Consider the equilibrium represented by

$$PbSO_4(s) + 3OAc^- \rightleftharpoons Pb(OAc)_3^- + SO_4^{--}$$

(a) Given that K_{sp} of $PbSO_4$ is 6.3×10^{-7} M^2 and that β_3, the overall formation constant of the complex, is 3×10^3 M^{-3}, calculate the equilibrium constant for the above reaction.

(b) Estimate the solubility of lead sulfate in 2 M ammonium acetate. Why is acetic acid unsuitable as a source of acetate?

8.21. It is found experimentally that cadmium sulfide will precipitate from a 0.010 M solution of cadmium nitrate upon saturating it with hydrogen sulfide at 25°C and 1 atm if the pH is greater than 0.22. Calculate the solubility product of active cadmium sulfide.

NINE

Cation Group 3. The Basic Hydrogen Sulfide Group: Al^{3+}, Cr^{3+}, Fe^{3+}, Fe^{++}, Mn^{++}, Co^{++}, Ni^{++}, Zn^{++}

9.1 Introduction

The eight ions of this group are precipitated as sulfides or hydroxides by a combination of hydrogen sulfide or thioacetamide with an ammonia–ammonium chloride buffer. None of these ions forms a chloride or sulfide sufficiently insoluble to precipitate from a solution that is 0.1–0.3 M in hydrochloric acid. The position of the seven elements that we study and of the less common members of Cation Group 3 are shown below. Heavy lines enclose the elements that are usually precipitated as sulfides; the rest are brought down as hydroxides or hydrated oxides. Vanadium and tungsten are, at best, only incompletely precipitated.

IA	IIA												IIIA
	Be												
		IIIB	IVB	VB	VIB	VIIB		VIII		IB	IIB	Al^{III}	
		Sc	Ti	(V)	Cr^{III}	Mn^{II}	Fe^{II}	Co^{II}	Ni^{II}		Zn^{II}	Ga	
		Y	Zr	Nb	Mo							In	
		La*	Hf	Ta	(W)							Tl	
		Ac†		* The lanthanides, atomic numbers 58–71.									
				† The actinides, atomic numbers 90–103.									

9.2   Structures of the Ions

The aluminum atom has a simple electronic structure that sets it apart from atoms of the other elements in this group: $1s^2 2s^2 2p^6 3s^2 3p^1$. The three valence electrons are normally lost to give Al^{3+} ion, exposing the electron configuration $1s^2 2s^2 2p^6$ found in the noble gas neon. The large expenditure of energy required for removal of three electrons, 1227 kcal mol^{-1}, is more than repaid by release of energy in hydrating the ion in solution or in combining it with anions such as sulfate in crystals. In some combinations, such as aluminum chloride, the bonds have considerable covalent character.

Table 9.1

The Electronic Structures of Atoms and Ions of Elements 20 to 30

		Electronic configurations*		
	Atomic			
Element	number	Atom, M	M^{++} ion	M^{3+} ion
Ca	20	$4s^2$	$4s^0$	—
Sc	21	$3d^1 4s^2$	—	$3d^0 4s^0$
Ti	22	$3d^2 4s^2$	$3d^2$	$3d^1$
V	23	$3d^3 4s^2$	$3d^3$	$3d^2$
Cr	24	$3d^5 4s^1$	$3d^4$	$3d^3$
Mn	25	$3d^5 4s^2$	$3d^5$	$3d^4$
Fe	26	$3d^6 4s^2$	$3d^6$	$3d^5$
Co	27	$3d^7 4s^2$	$3d^7$	$3d^6$
Ni	28	$3d^8 4s^2$	$3d^8$	—
Cu	29	$3d^{10} 4s^1$	$3d^9$	—
Zn	30	$3d^{10} 4s^2$	$3d^{10}$	—

* All the atoms contain an argon core: $1s^2 2s^2 2p^6 3s^2 3p^6$. The configuration $4s^0$ or $3d^0 4s^0$, denoting empty subshells, therefore indicates an ion with the noble gas configuration of argon.

The remaining elements of this group are in the first long period, and their atoms and ions differ in the number of $3d$ electrons as shown in Table 9.1. The $4s$ electrons are always lost first by the atom. The loss of another electron is frequently possible because the energies of the $3d$ and $4s$ states are not greatly different. All but zinc and scandium, therefore, give compounds in more than one oxidation state, the most common states being 2+ and 3+ (Table 9.2). In contrast with Al^{3+}, which may be called a d^0 ion, the d^n ions of the transition metals have distinctive properties: in particular, they are colored and attracted into a magnetic field (paramagnetic).

Table 9.2

Common Ions, Oxides, and Hydroxides of Cation Group 3*

Oxidation state	Al	Cr	Mn	Fe	Co	Ni	Zn
2+		(Cr^{++}) $[Cr(OH)_2]$	Mn^{++} $Mn(OH)_2$	Fe^{++} $Fe(OH)_2$ $Fe(CN)_6^{4-}$	Co^{++} $Co(OH)_2$ $Co(NH_3)_6^{++}$	Ni^{++} $Ni(OH)_2$ $Ni(NH_3)_6^{++}$	Zn^{++} $Zn(OH)_2$ $Zn(NH_3)_4^{++}$ $Zn(OH)_4^{--}$
3+	Al^{3+} Al_2O_3 $Al(OH)_3$ $Al(OH)_4^-$	Cr^{3+} Cr_2O_3 $Cr(OH)_4^-$ $[Cr(NH_3)_6^{3+}]$ $[Cr(CN)_6^{3-}]$	(Mn^{3+}) $MnO(OH)$	Fe^{3+} $FeO(OH)$ Fe_2O_3 $Fe(CN)_6^{3-}$	(Co^{3+}) $[Co_2O_3]$ $[Co(OH)_3]$ $Co(NH_3)_6^{3+}$ $Co(NO_2)_6^{3-}$ $Co(CN)_6^{3-}$	$NiO(OH)$	
4+			MnO_2				
6+		CrO_4^{--} $Cr_2O_7^{--}$ CrO_3	$[MnO_4^{-}]$				
7+			MnO_4^- $[Mn_2O_7]$				

* Some less common forms are given in parentheses or brackets for comparison. Rare oxidation states found only in a few complexes are not included.

The d^n ions have higher nuclear charges than the d^0 ions K$^+$ and Ca^{++} that precede them in the first long period. This higher nuclear charge is not effectively screened by the $3d$ electrons, for they do not penetrate as closely to the nucleus as the s and p eletrons do. The transition metal ions are therefore smaller and exert a stronger polarizing action on other ions and molecules than K$^+$ and Ca^{++} do. Complex formation is another characteristic property of these d^n ions.

9.3 Some Characteristics of the 2+ Oxidation State

We shall consider the Mn^{++}, Fe^{++}, Co^{++}, Ni^{++}, and Zn^{++} ions and shall add for comparison the Cu^{++} ion but shall exclude the comparatively uncommon Cr^{++} ion. The properties of the ions are summarized in Table 9.3. All have the same charge and about the same radius, so that many characteristics are similar:

(1) All form soluble chlorides, bromides, iodides, acetates, nitrates, sulfates, perchlorates, and thiocyanates.

(2) All form insoluble hydroxides, sulfides, carbonates, oxalates, cyanides, and phosphates. These are either polarizable or highly charged anions.

Table 9.3

Some Properties of the Bivalent Ions

Ion	Mn^{++}	Fe^{++}	Co^{++}	Ni^{++}	Cu^{++}	Zn^{++}
Atomic number	25	26	27	28	29	30
Ion radius, Å	0.80	0.75	0.72	0.69	0.72	0.74
log K for MS*	12.6	17.2	24.7	25.7	35.2	24.1
log K for M(OH)$_2$*	12.8	15.1	15.7	17.2	19.7	16.9
log K for M(NH$_3$)$_4$$^{++}$	—	3.7	5.0	7.5	12.0	8.7
log K for M(gl)$_2$†	6.0	7.8	8.9	11.0	15.4	9.3
log K for M(EDTA)$^{--\ddagger}$	13.6	14.3	16.2	18.6	18.8	16.3
Ionization energy, kcal mol^{-1}	532	555	574	594	646	631
$E°$ for M^{++}+2e$^-$ → M(s), V	-1.18	-0.44	-0.28	-0.25	$+0.34$	-0.76

* The equilibrium constants are for the most insoluble forms of the sulfides and hydroxides.
† The abbreviation gl$^-$ is used for the glycinate ion, C$_2$H$_4$O$_2$N$^-$.
‡ The abbreviation EDTA^{4-} is used for the ethylenediaminetetraacetate ion.

(3) They form several series of similar crystalline salts of which we mention only the vitriols; for example, red CoSO$_4$·7H$_2$O, pink MnSO$_4$·5H$_2$O, green FeSO$_4$·7H$_2$O, emerald green NiSO$_4$·7H$_2$O, and blue CuSO$_4$·5H$_2$O.

(4) All the cations hydrolyze slightly to give weakly acidic solutions:

$$M^{++} + H_2O \rightleftharpoons MOH^+ + H^+ \qquad K = 10^{-8} \text{ to } 10^{-10} \, M$$

On the other hand, there are subtle and important differences between the sulfides, hydroxides, and complex ions of the 2+ oxidation state. Table 9.3 gives values of log K for the formation of the precipitates and complexes, such as

$$M^{++} + 2OH^- \leftrightharpoons M(OH)_2(s)$$

$$M^{++} + 4NH_3 \leftrightharpoons M(NH_3)_4{}^{++}$$

The values of log K, and likewise of the equilibrium constants themselves, increase from Mn^{++} to Cu^{++} and then drop again for Zn^{++}. Thus Cu^{++} ion forms the least soluble hydroxide and sulfide and the most stable complexes. Because of the insolubility of the sulfide, copper comes down in Cation Group 2. The order of equilibrium constants runs counter to that of the ionic radii and parallel with the energies required to remove two electrons from the isolated atoms. This ionization energy is an indication of the attraction of the cation for electrons and is greatest for copper, so Cu^{++} ion would be expected to attract electron pairs of ligands most strongly. The standard electrode potentials at the bottom of the table show an analogous trend; copper is the least active metal, Cu^{++} ion the most easily reduced cation of the set.

9.4   Some Characteristics of the 3+ Oxidation State

The compounds in this state are less ionic than those of the bivalent state. The tervalent ions M^{3+} are smaller and more highly charged than the M^{++} ions (Table 9.4). Their electrostatic effect on anions and ligands is therefore greater. The tervalent ions form many complexes. Indeed, Cr(III) and Co(III), together with Pt(II) and Pt(IV), form more complexes than any other elements. If Al^{3+} falls behind the others in this respect, it is because of its d^0 structure.

Table 9.4

Some Properties of the Tervalent Ions

Ion	Cr^{3+}	Mn^{3+}	Fe^{3+}	Co^{3+}	Al^{3+}
Atomic number	24	25	26	27	13
Ionic radius, Å	0.64	0.70	0.67	0.64	0.50
K_h, M	1.6×10^{-4}	?	6.7×10^{-3}	?	1.1×10^{-5}
K_{sp} of M(OH)$_3$(s)	10^{-30}	?	10^{-37}	10^{-43}	10^{-32}

Some further generalizations can be made.

(1) All these ions form soluble acetates, nitrates, perchlorates, chlorides, bromides, iodides, thiocyanates, and sulfates.

(2) All form insoluble phosphates and hydroxides.

(3) They form several series of crystalline compounds, of which the alums are the most famous; for example, $KAl(SO_4)_2 \cdot 12H_2O$, $NH_4Fe(SO_4)_2 \cdot 12H_2O$, and $KCr(SO_4)_2 \cdot 12H_2O$.

(4) They are all more extensively hydrolyzed than the bivalent ions:

$$M^{3+} + H_2O \leftrightharpoons MOH^{++} + H^+ \qquad K_h$$

Values of K_h are given in Table 9.4. Because of the strongly acidic nature of these ions, their salts with weak acids are usually decomposed by water. The following are typical reactions of the tervalent ion with strongly basic anions:

$$2Al^{3+} + 3S^{--} + 6H_2O \rightarrow 2Al(OH)_3(s) + 3H_2S(g)$$

$$2Fe^{3+} + 3CO_3^{--} + 3H_2O \rightarrow 2Fe(OH)_3(s) + 3CO_2(g)$$

9.5 Preliminary Tests for the Ions

Certain tests can be made on the original sample if the test is specific for the particular ion and interfering ions are removed or masked. We consider first the identification reactions and then the methods of overcoming interferences.

Complex formation is the most commonly used identification. Iron(II) forms a blood red thiocyanate complex, and cobalt(II) forms a blue one:

$$Fe^{3+} + SCN^- \leftrightharpoons FeSCN^{++}$$

$$Co^{++} + 4SCN^- \leftrightharpoons Co(SCN)_4^{--}$$

Chelation is a special type of complex formation in which one ligand, acting like

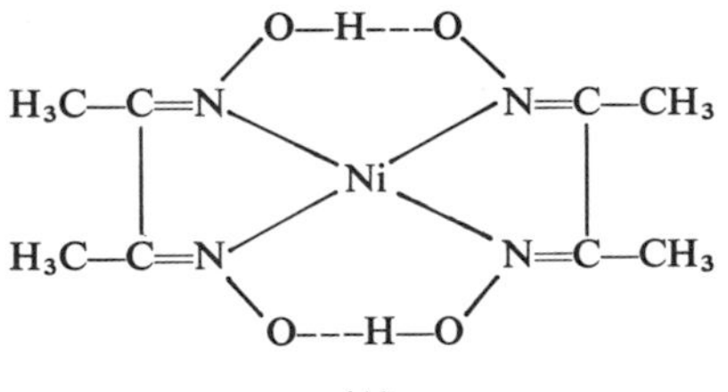

(A)

a crab's claws (Greek *chelos*), grasps the metal ion in more than one coordinating position. Dimethylglyoxime is used to detect nickel(II) by formation of a scarlet precipitate of the chelate (A) and the dithizone anion combines with zinc to give a

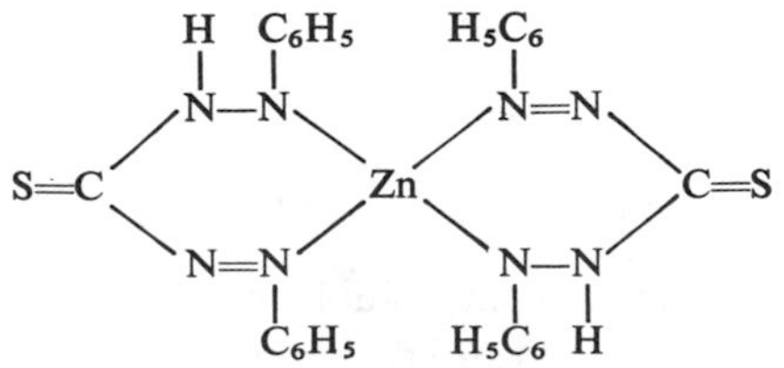

(B)

purple red chelate with structure (B). The reaction is represented by

$$Zn(OH)_4{}^{--} + 2C_{13}H_{11}N_4S^- \rightarrow Zn(C_{13}H_{11}N_4S)_2 + 4OH^-$$

An interesting coordination compound, Prussian blue, is used to identify Fe^{++} ion:

$$K^+ + Fe^{++} + Fe(CN)_6{}^{3-} \rightarrow KFeFe(CN)_6(s)$$

Manganese(II) ion is identified by oxidation to purple permanganate ion with "sodium bismuthate" (active principle Bi_2O_5). The complete equation can be derived from the following skeleton by the ion–electron method:

$$Mn^{++} + Bi_2O_5(s) \rightarrow MnO_4{}^- + 2Bi^{3+} \quad \text{(acid solution)}$$

Several different methods of overcoming interferences are used in the preliminary tests.

(1) *Precipitation*, as in the Ni and Mn tests. Insoluble hydroxides a reprecipitated and removed before addition of dimethylglyoxime lest they obscure a small scarlet precipitate of nickel dimethylglyoxime. Chlorides are removed by precipitation as silver chloride; they reduce bismuthate and permanganate and interfere with the test for Mn.

(2) *Masking or complexing*, as in the test for Co. Iron(III) forms a deep red complex with thiocyanate that can obscure the blue color of the cobalt thiocyanate. The addition of fluoride converts iron(III) to a weakly dissociated and colorless fluoride complex.

(3) *Redox*, as in the test for Co. Copper(II) and iron(III) are reduced by tin(II) chloride to copper(I) and iron(II), which give colorless compounds with thiocyanate.

(4) *Extraction*, as in the test for Co. Blue cobalt(II) thiocyanate complexes are soluble in a mixture of amyl alcohol and ether. Extraction into this solvent mixture serves to intensify the color and separate the complex from other colored ions such as complexes of nickel.

(5) *Differential diffusion and adsorption on spot paper*, as in the test for Zn. Ions differ in the extent to which they are adsorbed by paper and in their rates of diffusion through the solvent in the pores of the paper. In the zinc test, an alkaline solution of the sample is spotted on paper impregnated with dithizone. Hydroxide ions diffuse most rapidly and give an orange outer ring. Zincate ions are the next most active and give a purple red ring. The other ions diffuse so slowly or are adsorbed so tightly that they do not move at all, and they give a central spot.

9.6 Precipitation of the Group

A combination of ammonia, ammonium chloride, and thioacetamide or hydrogen sulfide is used to precipitate the group. The sulfides of Mn(II), Fe(II), Co(II), Ni(II), and Zn(II) have larger solubility products than the sulfides of Cation Group 2 and therefore require a higher concentration of sulfide ion for precipitation. The relation between acidity and the sulfide ion concentration has been discussed in Secs. 2.4 and 2.6. If the sample being analyzed for Group 3 has been treated previously with thioacetamide or hydrogen sulfide, iron will be in the form of Fe^{++} ions,

$$2Fe^{3+} + H_2S \rightarrow 2Fe^{++} + S + 2H^+$$

and will precipitate as FeS.

Aluminum and chromium(III) hydroxides have very low solubility products (Table 9.4). Even with the low concentration of hydroxide ion supplied by an ammonia–ammonium chloride buffer, they are almost completely precipitated, whereas magnesium hydroxide, which has a much larger solubility product, is not precipitated. A fuller account of this separation has been given in Secs. 2.7 and 2.8.

Some typical equations for the group precipitation are:

$$Mn^{++} + S^{--} \rightarrow MnS(s)$$

$$Co(NH_3)_6^{++} + S^{--} \rightarrow CoS(s) + 6NH_3$$

$$Al^{3+} + 3NH_3 + 3H_2O \rightarrow Al(OH)_3(s) + 3NH_4^+$$

9.7 Separation and Identification of Manganese

Manganese must be removed at an early stage in the analysis of the group, or it will cause trouble in later separations. The group precipitate is dissolved in nitric acid, which acts as an acid on the basic hydroxides of aluminum and chromium(III)—for example,

$$Cr(OH)_3(s) + 3H^+ \rightarrow Cr^{3+} + 3H_2O$$

and dissolves the sulfides by oxidizing sulfide to sulfur. Equations for the oxidation of the sulfides can be obtained by balancing such skeletons as

$$CoS(s) + NO_3^- \rightarrow NO_2(g) + Co^{++} + S(s)$$

by the ion–electron method. Hydrochloric acid is not suitable for dissolving the sulfides, because aged precipitates of CoS and NiS are not very soluble in it and because the presence of chloride is undesirable in the subsequent separation of manganese.

Manganese is isolated by oxidation to the dioxide with chlorate in 16 M nitric acid solution. The presence of chloride is undesirable because it reduces MnO_2, and chloride must be removed prior to the separation by repeated evaporation with nitric acid. Chromium(III) is also oxidized by chlorate to dichromate, and this is advantageous in later steps. Equations for these redox reactions can be derived by the ion–electron method from the following skeletons:

$$Cl^- + NO_3^- \rightarrow NOCl(g) + Cl_2(g)$$

$$Mn^{++} + ClO_3^- \rightarrow MnO_2(s) + ClO_2(g)$$

$$Cr^{3+} + ClO_3^- \rightarrow Cr_2O_7^{--} + ClO_2(g)$$

After separating the manganese dioxide, it is dissolved in acidified hydrogen peroxide, for which the skeleton is

$$MnO_2(s) + H_2O_2 \rightarrow O_2(g) + Mn^{++}$$

Manganese is then identified by oxidation to purple permanganate with sodium bismuthate (Sec. 9.5).

9.8 Subdivision of the Group

After removal of manganese, it is convenient to separate the amphoteric hydroxides from the basic ones. The amphoteric hydroxides of aluminum and zinc not only can be dissolved in acids but also are attacked by strong bases:

$$Al(OH)_3(s) + OH^- \rightarrow Al(OH)_4^-$$

$$Zn(OH)_2(s) + 2OH^- \rightarrow Zn(OH)_4^{--}$$

At the same time, dichromate is converted to chromate:

$$2OH^- + Cr_2O_7^{--} \rightarrow 2CrO_4^{--} + H_2O$$

The hydroxides of iron(III) and nickel are not appreciably affected, but $Co(OH)_2$ is somewhat soluble in excess base. Hydrogen peroxide is thus used in conjunction with the base to convert $Co(OH)_2$ to less soluble $Co(OH)_3$; the skeleton is

$$Co(OH)_2(s) + HO_2^- \rightarrow Co(OH)_3(s) + OH^-$$

The presence of manganese(II) or chromium(III) at this stage would lead to an unsatisfactory separation of zinc, for combinations of zinc and manganese or zinc and chromium coprecipitate with the hydroxides. Manganese was thus oxidized to the dioxide and chromium to dichromate in the preceding stage of analysis.

9.9 Identification of Iron, Cobalt, and Nickel

The basic hydroxides of iron, cobalt, and nickel are dissolved in hydrochloric acid. Hydrogen ions of the acid combine with hydroxide ion of the base, reducing the hydroxide ion concentration and liberating the metal ion

$$Fe(OH)_3(s) + 3H^+ \rightarrow Fe^{3+} + 3H_2O \qquad [Fe^{3+}][OH^-]^3 < K_{sp}$$

It should be noted that, whatever the oxidation state of iron in the original sample, it will be $3+$ here. When FeS in the group precipitate was dissolved in nitric acid, both iron and sulfur were oxidized. Cobalt(III) ion is unstable and reverts to Co^{++} ion, liberating oxygen from water. The equation can be derived from the skeleton

$$Co^{3+} + H_2O \rightarrow O_2 + Co^{++}$$

by the ion–electron method.

The three ions Fe^{3+}, Co^{++}, and Ni^{++} are then identified as in the preliminary tests (Sec. 9.5). Cobalt forms a brown complex with dimethylglyoxime (DMG^-), and sufficient reagent must be added to sequester the cobalt before the nickel chelate will precipitate:

$$Co(NH_3)_6{}^{++} + 3DMG^- \rightarrow Co(DMG)_3{}^- + 6NH_3$$

9.10   Identification of Chromium

The presence of chromate is generally indicated by a yellow color in the alkaline solution after removal of the hydroxides of iron, cobalt, and nickel. Further confirmation is obtained by the precipitation of bright yellow lead chromate from weakly acidic solution:

$$Pb(OAc)_2 + CrO_4{}^{--} \rightarrow PbCrO_4(s) + 2OAc^-$$

Additional confirmation is obtained by dissolving the precipitate in acid (what would the products be?) and forming blue, unstable CrO_5, a peroxide of chromium. The equation can be derived from the skeleton

$$Cr_2O_7{}^{--} + H_2O_2 \rightarrow CrO_5 + H_2O$$

The chromate–dichromate relationship is discussed further in Sec. 9.13.

9.11   Separation and Identification of Aluminum

Aluminum is separated from zinc and chromium by precipitation of the hydroxide from an ammonia–ammonium chloride buffer (Sec. 2.8). Zinc ions are sequestered in the weakly dissociated complex $Zn(NH_3)_4{}^{++}$, so that zinc hydroxide cannot precipitate (Sec. 2.11). The aluminum hydroxide is redissolved in acid, and aluminum is reprecipitated in the presence of aluminon reagent (a salt of aurintricarboxylic acid) to give the red chelate $Al(C_{22}H_{13}O_9)_3$. The pH must be regulated between 5 and 7.2. In more acidic solutions no reaction occurs; in more basic solutions the reagent itself gives a red color. The solution is therefore buffered with acetic acid and ammonium acetate. Iron(III) and chromium(III) interfere by giving colored reactions, but they should be absent if the preceding separations were carefully done. The principal purpose of this test is to distinguish between aluminum hydroxide and silica (hydrated SiO_2), both of which are gelatinous white precipitates. Silica is frequently introduced unwittingly by the use of alkaline reagents that become contaminated with it when stored in glass bottles. It does not give a red lake or color with aluminon.

9.12 Identification of Zinc

Zinc is identified first by precipitation of white zinc sulfide and then by the dithizone spot test (Sec. 9.5). Chromate must be removed by precipitation of barium chromate before precipitation of the sulfide, or it will oxidize thioacetamide or hydrogen sulfide to white colloidal sulfur, which might be confused with zinc sulfide. Zinc sulfide, unlike sulfur, is soluble in acids. Nitric acid is used to dissolve it in preference to hydrochloric acid in order to destroy the sulfide. Zinc must be converted to zincate ion, $Zn(OH)_4^{--}$, for the dithizone test, and if sulfide or hydrogen sulfide were present, ZnS would be reprecipitated upon addition of base and the dithizone test would fail.

Equations for the sequence of reactions involved here can be derived by completing and balancing the following:

$$Ba^{++} + CrO_4^{--} \rightarrow ?$$

$$Zn(NH_3)_4^{++} + S^{--} \rightarrow ?$$

$$ZnS(s) + NO_3^- \rightarrow Zn^{++} + NO_2(g) + S(s)$$

$$Zn^{++} + OH^- \rightarrow ?$$

The reaction between zincate and dithizone is discussed in Sec. 9.5.

9.13 Some Numerical Illustrations

A number of computations were done in Secs. 2.6, 2.8, and 2.11 that pertain to the analysis of this group. Some problems analogous to these were included in the exercises at the end of Chapter 2 and more are at the end of the present chapter.

(a) The Chromate–Dichromate Relationship

Chromium is identified by precipitating lead chromate in an acid solution. Under these conditions the concentration of chromate ion is very low, owing to two equilibria[1]:

$$2CrO_4^{--} + 2H^+ \leftrightharpoons Cr_2O_7^{--} + H_2O \qquad \beta_{22} = \frac{[Cr_2O_7^{--}]}{[H^+]^2[CrO_4^{--}]^2} = 10^{14.0}$$

$$(9.1)$$

$$CrO_4^{--} + H^+ \leftrightharpoons HCrO_4^- \qquad K_1 = \frac{[HCrO_4^-]}{[H^+][CrO_4^{--}]} = 10^{6.0}$$

$$(9.2)$$

---

[1] Values of these equilibrium quotients are sensitive to interionic forces. Because $[CrO_4^{--}]$ is low, we fortunately need only approximate values for the computation.

Chromium exists in three forms in the solution: CrO_4^{--}, $HCrO_4^-$, and $Cr_2O_7^{--}$ ions. The total chromium in moles per liter must be given by

$$C_T = [CrO_4^{--}] + [HCrO_4^-] + 2[Cr_2O_7^{--}] \qquad (9.3)$$

where we have to count each Cr of $Cr_2O_7^{--}$ so the concentration of the ion is doubled. We take $[CrO_4^{--}]$ as a convenient unknown because it is small in the acid solution. By combining equations (9.1)–(9.3), we obtain

$$C_T = [CrO_4^{--}] + K_1[CrO_4^{--}][H^+] + 2\beta_{22}[CrO_4^{--}]^2[H^+]^2$$

In the analysis of the practice solution for this group we start with a sample containing 0.5 mg of chromium; by the time the identification stage is reached, one-fourth of this will be in a volume of roughly 1 ml, whence

$$C_T = \frac{0.125 \text{ mg ml}^{-1}}{52 \text{ mg mmol}^{-1}} = 2.5 \times 10^{-3} \ M$$

A slight excess of acetic acid is present, so the pH is about 3, making $[H^+] = 10^{-3} \ M$. The equation we have to solve is therefore

$$2.5 \times 10^{-3} = [CrO_4^{--}](1 + 10^6 \times 10^{-3}) + 2 \times 10^{14} \times 10^{-6}[CrO_4^{--}]^2$$
$$= 1001[CrO_4^{--}] + 2 \times 10^8 \times [CrO_4^{--}]^2$$

Putting $x = [CrO_4^{--}]$ we can reduce this to

$$x^2 + 5 \times 10^{-6}x - 1.25 \times 10^{-11} = 0$$

which can be solved by the quadratic formula

$$x = \frac{-5 \times 10^{-6} \pm (25 \times 10^{-12} + 50 \times 10^{-12})^{1/2}}{2} = 1.8 \times 10^{-6} \ M$$

In view of the fact that the solubility product of lead chromate is $2 \times 10^{-16} \ M^2$, we need a lead ion concentration larger than $(2 \times 10^{-16})/(1.8 \times 10^{-6})$, or about $10^{-10} \ M$, to precipitate lead chromate. Any soluble lead salt, even a weakly dissociated one like lead acetate, is a satisfactory source of lead ion for this purpose.

(b) Oxidation of Cobalt(II)

The comparative ease of oxidation of cobalt(II) to cobalt(III) can be deduced from the standard electrode potentials
Acid solution:

$$Co^{3+} + e^- \rightarrow Co^{++} \qquad E° = 1.84 \text{ V}$$

$$2H^+ + H_2O_2 + 2e^- \rightarrow 2H_2O \qquad E° = 1.77 \text{ V}$$

Basic solution:

$$Co(OH)_3(s) + e^- \rightarrow Co(OH)_2(s) + OH^- \qquad E° = 0.17 \text{ V}$$

$$H_2O + HO_2^- + 2e \rightarrow 3OH^- \qquad E° = 0.88 \text{ V}$$

In acid solution, E° for the cobalt couple is greater than that of the oxidizing agent, hydrogen peroxide, so we do not expect oxidation to occur to an appreciable extent. The reverse is true in basic solution, and in the analytical procedure cobalt(II) is oxidized in the presence of excess sodium hydroxide.

To give the comparison more quantitative meaning, we can calculate E° and K for the overall reactions

Acid solution:

$$2Co^{++} + H_2O_2 + 2H^+ \rightleftharpoons 2H_2O + 2Co^{3+}$$

$$E^\circ = 1.77 \text{ V} - 1.84 \text{ V} = -0.07 \text{ V}$$

$$\log K = \frac{E^\circ}{k_N/n} = \frac{-0.07}{0.0592/2} = -2.4$$

$$K = 4 \times 10^{-3} \, M^{-3}$$

Basic solution:

$$2Co(OH)_2(s) + H_2O + HO_2^- \rightleftharpoons 2Co(OH)_3(s) + OH^-$$

$$E^\circ = 0.88 \text{ V} - 0.17 \text{ V} = 0.71 \text{ V}$$

$$\log K = \frac{0.71}{0.0592/2} = 24.0$$

$$K = 1 \times 10^{24}$$

The large value of the latter constant indicates again the desirability of using basic solutions for the oxidation.

EXPERIMENTAL PART

9.14 Analysis of Known and Unknown Samples[2]

(a) Use 5 drops of a *known* or *practice solution* that contains 2 mg/ml each of Al^{3+}, Cr^{3+}, Ni^{++}, Co^{++}, and Mn^{++} and 4 mg/ml of Zn^{++}. *Note that the practice solution contains no iron.* Carry out the tests for iron and note the faint colorations that may appear owing to impurities. Then add one drop of Fe^{3+} solution to the test and observe the appearance of a bona fide test.

(b) If the *unknown* is Solution 2.1 from Outline 2, use all of it in the systematic analysis. If it is an unknown solution containing only the ions of this group, use 5 drops for the systematic analysis; more must be available for the preliminary tests. If the unknown is a solid, take a representative sample (T-1) and dissolve about 50 mg in water or 6 M HCl; use about a third of this for the systematic analysis.

[2] The absence of phosphate, oxalate, tartrate, and other interfering anions is assumed.

Always note the color of the known or unknown samples. Compare with the test solutions of the ions. Some of the most common ions and colors are: Cr^{3+}, violet; $Ni(NH_3)_6^{++}$, blue; $CoCl_2$ and $CoCl_4^{--}$, blue; $CrCl_2^{+}$ and $Cr(OH)_4^{-}$, deep green; Fe^{++}, pale green; $FeCl^{++}$, yellow; Fe(III)—OH complexes, yellow to red; Co^{++}, red; Mn^{++}, very pale pink, almost colorless; Al^{3+} and Zn^{++}, colorless. In the solid state, combinations of colors must be considered: green nickel and red cobalt salts give a gray mixture.

9.15   Preliminary Tests

It is possible to make specific tests for all of the ions of this group except aluminum and chromium on the original sample, even when it contains ions of other groups. These tests are given in PT-1 through PT-5. It is wise to confirm them by the systematic analysis according to Outline 5.

PT-1.   Tests for Fe

In a general analysis all Fe^{3+} ions are reduced to Fe^{++} by H_2S or TA. Tests for the oxidation state of iron must be made on the original sample.

(a) *Test for* Fe^{++}. Dissolve one small crystal of $K_3Fe(CN)_6$ in about 20 drops of water. Add a drop of this solution to a drop of the solution under test. A blue color or blue precipitate of $KFeFe(CN)_6$ indicates the presence of Fe^{++}. If the test is green, the yellow color of the ferricyanide reagent may be too strong. Dilute it and try the test again. Other ions give colored precipitates with this reagent, but only Fe^{++} gives a blue one.

(b) *Test for* Fe^{3+}. Dissolve a few crystals of NH_4SCN in a little water, and add a few drops of this reagent to the solution under test. A dense red color, not a precipitate, of $FeSCN^{++}$ indicates the presence of Fe^{3+} ion. Light pink or red colorations are caused by traces of iron, a common impurity in reagents. All aged solutions of Fe^{++} will give a test for Fe^{3+}, because the iron(II) is slowly oxidized by oxygen of air. Do not report traces. If you are in doubt as to whether the color is deep enough, try the test on a few drops of a solution that contains 0.1 mg of Fe^{3+} per ml (dilute one drop of the standard ion solution with 5 ml of water and mix well). Indications of the presence of iron are also obtained in the test for cobalt (PT-4).

PT-2.   Test for Ni

To one drop of the solution under test, add a few drops of water and 15 M NH_3 until the solution is strongly alkaline. Centrifuge and draw off the clear solution to another tube. To it add several drops of dimethylglyoxime solution. A deep red precipitate of nickel dimethylglyoxime indicates the presence of Ni. If a brown coloration caused by Co forms, add more reagent. This test will fail when only a small amount of nickel is present because of coprecipitation with a large precipitate of the gelatinous hydroxides.

PT-3.   Test for Mn

Test a drop of the solution for chloride by acidifying the solution with nitric acid, if necessary, and adding $AgNO_3$ solution.

(*a*) If chloride is absent, add a drop of 16 *M* HNO_3 and one-quarter spatula-full of sodium bismuthate. A purple color of MnO_4^- indicates the presence of Mn. Centrifuge to see the color.

(*b*) If chloride is present, add $AgNO_3$ solution until no more precipitate forms. Then test with HNO_3 and sodium bismuthate as before.

PT-4.   Test for Co

To a drop of the (acidic) solution add a few drops of water and several crystals of NH_4SCN. If a red color ($FeSCN^{++}$) appears, add solid NaF a little at a time until the color goes away. Then add 5–10 drops of a solution of NH_4SCN in alcohol–ether. A blue upper layer containing $Co(SCN)_4^{--}$ indicates the presence of Co. A green upper layer indicates the presence of low concentrations of either Co^{++} or SCN^-. If the color is indistinct or fades on shaking, add a spatula-full of solid NH_4SCN. If the color is red, add more NaF. A dark muddy green color may be due to Cu; add a drop of $SnCl_2$ to reduce Cu(II) to CuSCN or $Cu(SCN)_2^-$.

PT-5.   Test for Zn

To one drop of the solution under test, add 6 *M* NaOH until the solution is basic, and then add 1 or 2 drops in excess. Touch a pipet to this mixture without squeezing the bulb. Some of the solution will rise in the tip by capillary action. Bring the tip down vertically on the center of a square of dithizone paper. A purple red spot indicates the presence of Zn. An orange ring is caused by NaOH alone. Mercury(I or II) and lead(II) give purple blue spots, and Sn(II) gives a pink one. Touch the pipet to some water, and bring the tip down on the center of the spot. The zinc spot will spread as the water flows out; the other spots will not.

OUTLINE 5. The Systematic Analysis of Cation Group 3

Precipitation of the Group: Dilute sample with water until total vol is about 2 ml. Add 6 drops 6 M NH_4Cl and make soln alk with 15 M NH_3 (check with litmus) and add 1 drop NH_3 in xs.[1] Note the occurrence and color of any ppt.[2] Then add 4 drops TA and heat in vigorously boiling water for at least 5 minutes. Centrifuge and wash ppt with hot water containing a drop of 0.2 M NH_4NO_3 soln.[3] Combine the washing with Soln 5.1.

Solution 5.1. Ions of later groups, NH_3, TA, NH_4Cl, H_2S. If Ppt 5.1 is large, check with litmus and, if necessary, add more NH_3 and TA. Reheat. If additional ppt appears, centrifuge and remove.

In general analyses, add 1 ml 12 M HCl and evap almost to dryness to remove H_2S and TA. Dilute with water, stopper, and save for **Group 4, Outline 6.**

Precipitate 5.1. $Al(OH)_3$, $Cr(OH)_3$, FeS, NiS, CoS, MnS, ZnS. Add 10 drops 16 M HNO_3 and warm until ppt dissolves.[4] Centrifuge and transfer soln to casserole. Discard the residue of S.

Solution 5.2. Fe^{3+}, Al^{3+}, Cr^{3+}, Ni^{++}, Co^{++}, Mn^{++}, Zn^{++}, HNO_3, (Cl^-). Evap almost to dryness over a microflame (T-7). Add 10–15 drops 16 M HNO_3 and manipulate the casserole so that it wets all of the inner wall. Evap almost to dryness again. Add more 16 M HNO_3 and evap a third time.[5] Add 10–15 drops 16 M HNO_3.

Heat soln to a boil and add to it a few crystals of $KClO_3$ at a time. Boil after each addn and add 16 M HNO_3 to replace that lost by evapn.[7] Continue the addn of $KClO_3$ until a brown black ppt forms[8] or until one-half spatula-full of $KClO_3$ (50–75 mg) has been added. Transfer to a centrifuge tube and rinse casserole with a few drops of water, adding the rinsings to the tube. Centrifuge and wash ppt with water.

A side test for Mn can be made on this soln if desired. Dilute 1 drop with several drops of water in a centrifuge tube. Add a definite xs of sodium bismuthate. Centrifuge.

Deep purple soln,[6] MnO_4^-: **presence of Mn.**

Precipitation 5.3. MnO_2. If previous test was doubtful, add a few drops 6 M HNO_3 and 1 drop 3% H_2O_2. Then add an xs of sodium bismuthate.[9] Centrifuge.

Purple soln, MnO_4^-: **presence of Mn.**

Solution 5.3. Fe^{3+}, Al^{3+}, $Cr_2O_7^{--}$, Ni^{++}, Co^{++}, Zn^{++}, HNO_3. Neutralize the acid with 6 M NaOH and add 5–10 drops in xs. Follow with a drop of 3% H_2O_2. Warm for a few minutes.[10] Centrifuge and wash ppt with about 10 drops H_2O containing 1–2 drops NaOH. Combine washings and soln in a centrifuge tube and centrifuge again to remove last trace of Ppt 5.4.[11]

Co and Fe tests. Cool.[12] Add a few crystals of NH_4SCN.

Deep red color, $FeSCN^{++}$: **presence of Fe.**[13]

Discharge color with NaF. Add 10 drops of soln of NH_4SCN in alcohol–ether.

Blue upper layer, $Co(SCN)_4^{--}$: **presence of Co.**[14]

Solution 5.4
(see facing page)

Precipitate 5.4
(see facing page)

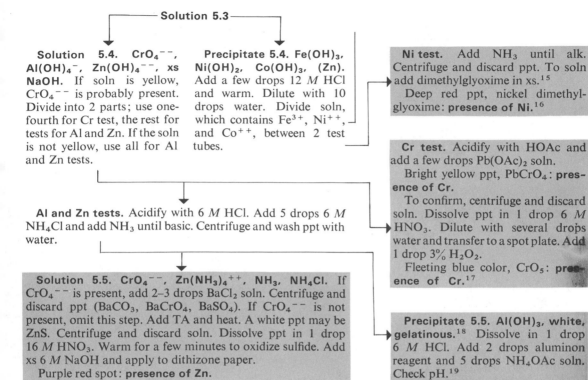

┌─────────────── **Solution 5.3** ───────────────┐

Solution 5.4. CrO_4^{--}, $Al(OH)_4^-$, $Zn(OH)_4^{--}$, xs NaOH. If soln is yellow, CrO_4^{--} is probably present. Divide into 2 parts; use one-fourth for Cr test, the rest for tests for Al and Zn. If the soln is not yellow, use all for Al and Zn tests.

Precipitate 5.4. $Fe(OH)_3$, $Ni(OH)_2$, $Co(OH)_3$, (Zn). Add a few drops 12 M HCl and warm. Dilute with 10 drops water. Divide soln, which contains Fe^{3+}, Ni^{++}, and Co^{++}, between 2 test tubes.

Ni test. Add NH_3 until alk. Centrifuge and discard ppt. To soln add dimethylglyoxime in xs.[15]
Deep red ppt, nickel dimethyl-glyoxime: **presence of Ni.**[16]

Cr test. Acidify with HOAc and add a few drops $Pb(OAc)_2$ soln.
Bright yellow ppt, $PbCrO_4$: **presence of Cr.**
To confirm, centrifuge and discard soln. Dissolve ppt in 1 drop 6 M HNO_3. Dilute with several drops water and transfer to a spot plate. Add 1 drop 3% H_2O_2.
Fleeting blue color, CrO_5: **presence of Cr.**[17]

Al and Zn tests. Acidify with 6 M HCl. Add 5 drops 6 M NH_4Cl and add NH_3 until basic. Centrifuge and wash ppt with water.

Solution 5.5. CrO_4^{--}, $Zn(NH_3)_4^{++}$, NH_3, NH_4Cl. If CrO_4^{--} is present, add 2–3 drops $BaCl_2$ soln. Centrifuge and discard ppt ($BaCO_3$, $BaCrO_4$, $BaSO_4$). If CrO_4^{--} is not present, omit this step. Add TA and heat. A white ppt may be ZnS. Centrifuge and discard soln. Dissolve ppt in 1 drop 16 M HNO_3. Warm for a few minutes to oxidize sulfide. Add xs 6 M NaOH and apply to dithizone paper.
Purple red spot: **presence of Zn.**

Precipitate 5.5. $Al(OH)_3$, white, gelatinous.[18] Dissolve in 1 drop 6 M HCl. Add 2 drops aluminon reagent and 5 drops NH_4OAc soln. Check pH.[19]
Red color: **presence of Al.** Make barely alk with NH_3 and centrifuge to settle the red lake.

Notes on Outline 5

1. The solution must be definitely basic, yet if it is too basic, small amounts of Al and Cr may be lost.

2. The precipitate may be white $Al(OH)_3$, gray green $Cr(OH)_3$, or reddish brown $Fe(OH)_3$ (this will not form if the iron has been reduced by TA or H_2S in acidic solution). Absence of these precipitates indicates probable absence of these ions, but the precipitates are sometimes hard to see when suspended in a colored solution.

3. Ammonium nitrate prevents the sulfides from going into colloidal suspension.

4. A black, floating globule of sulfur colored by occluded sulfides often forms. If difficulty is experienced in dissolving the sulfides, add a drop of 12 M HCl—but avoid this if possible, for chloride must be removed in the next step.

5. Repeated evaporation is necessary to remove chloride, which interferes with the separation of MnO_2.

6. If the purple color appears and fades rapidly, chloride is present. Add more bismuthate.

7. The oxidation of Mn(II) to MnO_2 by chlorate requires vigorous boiling, the presence of 16 M HNO_3, and the absence of chloride. Avoid adding too much $KClO_3$ at a time, or evolution of ClO_2 may be too violent.

8. If the side test shows the absence of Mn, no precipitate will be obtained. Even so, the potassium chlorate treatment is made to oxidize Cr(III) to $Cr_2O_7^{--}$.

9. The first bismuthate that is added will oxidize H_2O_2 to O_2. An excess is present when the layer of brown or black solid on the bottom is several millimeters thick and no bubbles are evolved.

10. The hydrogen peroxide oxidizes $Co(OH)_2$ to $Co(OH)_3$, and a better separation of Co from Al and Zn results. Warming serves to accelerate the reaction and decompose excess H_2O_2.

11. A double centrifugation is advisable to ensure complete separation of Fe, Co, and Ni from Cr, Al, and Zn.

12. The thiocyanate reagent contains ether and amyl alcohol, which are volatile solvents. If the solution is warm when the reagent is added, most of the solvents will evaporate.

13. The preliminary test PT-1 is a more reliable indicator of iron than a red color at this stage. A certain amount of iron frequently is picked up during the analysis from impurities in the reagents or through careless technique. Thus even when no iron is detected in PT-1, a pink color is sometimes obtained here.

14. If the blue color fades after mixing, add a spatulafull of solid NH_4SCN. A large water layer extracts so much thiocyanate from the upper layer that the sensitivity of the test is reduced.

15. A large, false nickel test may be found here if at any earlier stage the solution was stirred with a nickel spatula. Use glass rods for stirring.

16. If cobalt is present, it combines with dimethylglyoxime to give a brown *solution* before any dimethylgloyoxime will react with nickel. If a brown color forms, be sure to add a sufficient excess of reagent.

18. If the aluminum hydroxide is brownish, it may contain iron. Dissolve in a drop or two of 6 M NaOH and centrifuge to remove the brown $Fe(OH)_3$. Neutralize the solution with 6 M HCl, and add NH_4Cl and NH_3 to reprecipitate $Al(OH)_3$.

19. The best lake is formed if the pH of the acidic solution containing the aluminon is raised to between 5 and 7.2 by addition of the ammonium acetate. A precipitate may form in this weakly acidic solution if large amounts of Al are present. Otherwise, it will not appear until the solution is made weakly basic. An excess of NH_3 is harmful. Silica does not give a red lake, but Fe, Cr, Pb, and other metals do; they will not be present here if the earlier separations were carefully done.

EXERCISES

9.1. Name the elements of this group and locate them in the periodic table. Why do they occur together in this analytical group?

9.2. For each element in Group 3 give the principal oxidation states and the electronic structures of the simple ions.

9.3. What are some characteristic properties of d^n ions?

9.4. Name Cr_2O_3, $Fe(CN)_6^{4-}$, $Zn(OH)_4^{--}$, $Mn(OH)_2$, $Cr_2O_7^{--}$, MnO_4^{-}, $Cr(OH)_4^{-}$, $CoCl_2$, $Al(OH)_4^{-}$, $Co(NH_3)_6^{++}$.

9.5. Give the colors of Co^{++}, Cr^{3+}, Fe^{++}, CrO_4^{--}, $Ni(NH_3)_6^{++}$, MnS, $Cr_2O_7^{--}$, Ni^{++}, $Cr(OH)_3$, CoS, MnO_4^-.

9.6. Explain why the combination of ammonia and ammonium chloride does not precipitate $Mg(OH)_2$, $Mn(OH)_2$, and $Ni(OH)_2$. Reference should be made to appropriate equilibria. [Hint: There is more than one reason.]

9.7. Discuss the conditions chosen for the precipitation of the sulfides of Group 3, making reference to appropriate equilibria.

9.8. Why is Cr_2S_3 decomposed by water? Why does $Cr_2(SO_3)_3$ not precipitate when sodium sulfite is added to a solution of a chromium(III) salt? What precipitate is obtained?

9.9. Account for the observation that a pink color with thiocyanate is not obtained from a freshly prepared solution of iron(II) sulfate but does develop after the solution has stood for some time.

9.10. Fill in the spaces in the table with the formulas of the compounds or ions.

Ion	H_2S in 0.01 M HCl	H_2S in NaOAc–HOAc	H_2S in NH_3–NH_4Cl	Excess NaOH	Excess NH_3
Al^{3+}	Al^{3+}	Al^{3+}	$Al(OH)_3$		
Cr^{3+}	Cr^{3+}			$Cr(OH)_4^-$	
Mn^{++}	Mn^{++}				
Fe^{++}	Fe^{++}				
Fe^{3+}			FeS		
Co^{++}	Co^{++}			$Co(OH)_3^-$	
Ni^{++}	Ni^{++}				
Zn^{++}					

9.11. Suggest an error or errors that could lead to the following observations:
(a) In the manganese test the purple color fades quickly.
(b) In the nickel test only a dark brown solution was obtained although nickel was present.
(c) A pale green upper layer is obtained in the cobalt test although cobalt is present.

9.12. Give a reason for each operation:
(a) Evaporation with nitric acid before the separation of Mn.
(b) The use of hydrogen peroxide with sodium hydroxide in subdividing the group.
(c) Removal of chromate before the test for Zn.
(d) Separation of Mn before subdividing the group.
(e) The addition of sodium fluoride in the Co test.

9.13. Write equations for the successive reactions of each ion of this group.

9.14. Given the following samples, what is the briefest procedure you could use to answer the question posed in each part?

(a) Is a black pigment used in a Stone Age mural charcoal or pyrolusite (MnO_2)?

(b) Is a spatula made of nickel, nickel silver (Ni, Cu, Zn), or monel (Ni, Cu)?

(c) Are the yellow flakes you found in a sandbank gold or mica coated with fool's gold (FeS_2)?

(d) Does the white cream, which has lost its label, contain ZnO, an astringent and antiseptic, or $Al(OH)_2Cl$, an antiperspirant?

9.15. Devise the briefest possible procedures that could be used to analyze the following samples:

(a) A solution containing Cr^{3+}, Zn^{++}, and Al^{3+} ions.

(b) A solution of the pigment Thenard's blue containing Al^{3+} and Co^{++} ions.

(c) A solution of silver solder containing Ag^+, Cu^{++}, and Zn^{++} ions.

(d) A solution of Chromel thermocouple wire containing Cr^{3+}, Fe^{3+}, and Ni^{++} ions.

(e) The electrolyte of a used dry cell containing $Zn(NH_3)_4Cl_2$, NH_4Cl, and MnO_2.

9.16. Titanium is the seventh most abundant element in the earth's crust and seems to be even more abundant in moon rocks. In solution, titanium is usually found in an oxidation state of 4+, mainly as $Ti(OH)_3^+$ and $Ti(OH)_2^{++}$ ions. Its hydrated oxide, $TiO_2 \cdot xH_2O$, is less soluble than the oxides or hydroxides of aluminum, iron(III), and chromium(III) and is not amphoteric. Its sulfide cannot be precipitated from aqueous solution. Titanium (IV) does not give a colored complex with thiocyanate, but it reacts with hydrogen peroxide in acidic solution to give yellow to red TiO_2^{++}. How would you modify the systematic analysis so that titanium could be detected?

9.17. Given a 0.010 M solution of zinc chloride acidified with hydrochloric acid, what is the minimum concentration of acid necessary so that zinc sulfide will not precipitate when the solution is saturated with hydrogen sulfide at 25°C and 1 atm? Take the solubility product of active zinc sulfide to be $2.5 \times 10^{-22}\ M^2$.

9.18. A mixture of aged precipitates of cobalt(II) and iron(II) sulfides is to be separated by the action of hydrochloric acid. What must the concentration of acid be if no more than 0.010 mg of Co^{++} per ml is to dissolve? What concentration of iron(II) ion will be present in a solution of this acidity?

9.19. Calculate the minimum number of grams of ammonium nitrate that must be present in 2.5 ml of solution that also contains 0.50 mg of Mn^{++} ion, if no manganese(II) hydroxide is to precipitate when the solution is made 0.30 M in ammonia.

9.20. Calculate the equilibrium constant for the oxidation of hydrogen sulfide by iron(III) ion at 25°C, given

$$e^- + Fe^{3+} \rightarrow Fe^{++} \qquad E° = 0.771\ \text{V}$$

$$2e^- + S(s) + 2H^+ \rightarrow H_2S \qquad E° = 0.141\ \text{V}$$

TEN

Cation Group 4.

The Ammonium Carbonate

Group: Ca^{++}, Sr^{++}, Ba^{++}

10.1 Introduction

The three ions of this group form carbonates that are precipitated by ammonium carbonate in the presence of an ammonium chloride–ammonia buffer. Their chlorides, sulfides, and hydroxides are much too soluble to precipitate with preceding groups.

Calcium, strontium, and barium are the alkaline earth metals of periodic group IIA. Radium, the last member of this group, resembles barium in its chemical behavior but is not considered here because of its scarcity and radioactivity. Beryllium, the first member, resembles aluminum more closely than it

IA			
	IIA		
		IIIB	
	Ca^{II}		
	Sr^{II}		
	Ba^{II}		
	(Ra^{II})		

does the other members of Group IIA and accompanies aluminum in the analytical separations of qualitative analysis. Magnesium, too, stands apart from the alkaline earth metals, although it resembles calcium more closely than it does beryllium. By proper choice of conditions, little, if any magnesium carbonate is precipitated with Cation Group 4.

Atoms of these elements have two valence electrons of the s type and lose them to form M^{++} ions with electron configurations like those of the nearest noble gases. We call them d^0 ions because they have no outer-shell electrons in d orbitals. The formation of gaseous Ca^{++} ions from atoms, for example, requires the expenditure of 415 kcal mol^{-1}, a considerable amount of energy, but this investment is more than repaid by release in energy that occurs when stable crystals or hydrated ions are formed. Calcium, strontium, and barium always occur in crystals and in solution as bivalent ions.

10.2 Some Properties of the Ions

Some of the pertinent characteristics of the ions of Cation Group 4 are given in Table 10.1. Like all d^0 ions, these are colorless and hard to reduce to the free metal. Each is smaller than the alkali metal ion in the same period and has twice the charge. Their electrostatic effect, measured by the ratio of charge to radius, is considerably larger than that of the alkali metal ions but less than those of the transition metal ions that follow them in their periods. Nor do d^0 ions show the nonelectrovalent interaction so characteristic of ions like Cu^{++}, Ag^+, Hg^{++}, and Pb^{++}. Compounds of Cation Group 4 are typically ionic substances. Because of their double charges, these cations tend to form very stable lattices with small or highly charged anions, such as F^-, O^{--}, CO_3^{--}, or SO_4^{--}.

The solubility of a substance depends largely on a close balance between the energy required to take the crystal apart into separate, gaseous ions (the lattice energy in Table 10.1) and energy released in taking the gaseous ions into solution (the hydration energy in Table 10.1). It is not possible to calculate solubilities with

Table 10.1

Some Properties of the Ions of Cation Group 4

Property	*Ca^{++}*	*Sr^{++}*	*Ba^{++}*
Ionic radius, Å	0.99	1.14	1.35
Charge/radius ratio	2.0	1.8	1.5
Heat of hydration of M^{++}(g), kcal mol^{-1} released	380	345	311
Lattice energy of MCl_2, kcal mol^{-1} absorbed	537	506	488
Lattice energy of MF_2, kcal mol^{-1} absorbed	629	597	564
Lattice energy of MO, kcal mol^{-1} absorbed	823	781	741

any confidence from such energies (Sec. 1.7), but they give us some help in understanding common solubility rules. The solubilities of most compounds of Cation Group 4 vary regularly from calcium to barium, the exceptions being the carbonates, acetates, and perchlorates. Calcium compounds are generally the most soluble; this is true of the chlorides, bromides, iodides, chlorates, bromates, iodates, nitrites, nitrates, sulfites, sulfates, and chromates. For such compounds, solubility is controlled largely by the hydration energy, which makes Ca^{++} ion the most stable in solution. The anions are comparatively large, which makes the lattice energies comparatively low. For the hydroxides, oxalates, and fluorides—salts with small anions—barium compounds are the most soluble. Here lattice energy is the important effect; crystal lattices with the largest cation, Ba^{++}, are easiest to take apart and disperse in the solvent. The molar solubilities of some of the most important compounds are presented graphically in Fig. 10.1.

The effect of hydration is also seen in the formulas of the solid salts. Those of calcium are most frequently hydrated: $CaCl_2 \cdot 6H_2O$, $Ca(NO_3)_2 \cdot 4H_2O$, $CaSO_4 \cdot 2H_2O$, whereas the barium salts, because the cation is larger and attracts water dipoles less strongly, are less hydrated or even anhydrous: $BaCl_2 \cdot 2H_2O$, $Ba(NO_3)_2$, $BaSO_4$.

Large ions of the d^0 type, like those of Cation Group 4, have less tendency to form complexes than those of the preceding analytical groups. They coordinate best with ligands containing oxygen atoms, such as sugars and alcohols. They form chelates with triethanolamine and ethylenediaminetetraacetic acid (EDTA) in which the cation is bonded to both oxygen and nitrogen atoms. Calcium ion is the smallest of the three cations and forms the most stable complexes.

10.3 Precipitation of the Group

The elements of Cation Group 4 are so closely related, and their chemistry is so undistinguished—similar solubilities, few complex ions, no redox reactions of use in analysis—that separations are difficult.

All three cations form carbonates with similar solubilities in water (Fig. 10.1), so that it is convenient to precipitate the group as carbonates; for example,

$$Ca^{++} + CO_3^{--} \rightarrow CaCO_3(s)$$

Another advantage in precipitating the group as carbonates is that we can then get rid of this anion by the simple expedient of adding an acid; for example,

$$SrCO_3(s) + 2HOAc \rightarrow Sr^{++} + 2OAc^- + H_2O + CO_2(g)$$

From this solution we can then separate and identify Ba^{++}, Sr^{++}, and Ca^{++} in turn by making use of solubility differences and the greater ability of Ca^{++} ion to form complexes.

Some care is required in choosing the conditions for precipitation of the group because magnesium carbonate is also somewhat insoluble; its solubility product

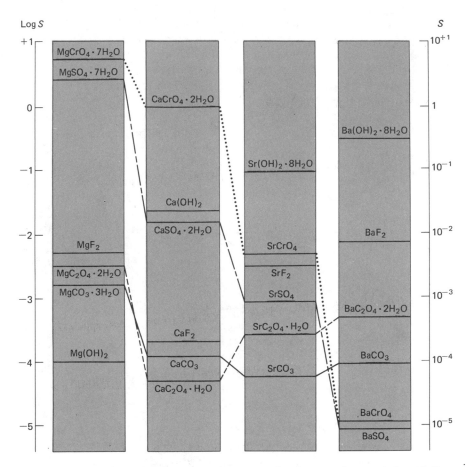

Fig. 10.1. Molar solubilities (S) of compounds of magnesium and the alkaline earth metals in water at 25°C.

is about 10^{-5} M^2 as compared with 10^{-9} for the three carbonates of Cation Group 4. By controlling the carbonate ion concentration (Sec. 2.9) we keep it low enough to establish conditions such as

$$[Mg^{++}][CO_3^{--}] < K_{sp} \quad \text{and} \quad [Ca^{++}][CO_3^{--}] > K_{sp}$$

The principal equilibrium in the buffer mixture of ammonium carbonate, ammonium chloride, and ammonia that is used to precipitate the group is expressed by

$$NH_3 + HCO_3^- \leftrightharpoons NH_4^+ + HCO_3^-$$

The equilibrium constant is 0.082 at 25°C and 0.023 at 50°C and can be written in the form

$$\frac{[CO_3^{--}]}{[HCO_3^{-}]} = K \times \frac{[NH_3]}{[NH_4^{+}]} \tag{10.1}$$

By using roughly equal concentrations of ammonia and ammonium ion, we can keep the proportion of free carbonate ion low because K is small. Then magnesium carbonate will not precipitate. Without addition of ammonium ion from ammonium chloride, the proportion of free carbonate might be too high. On the other hand, if too much ammonium ion is present, the concentration of carbonate could be so reduced that Group 4 would not be completely precipitated. This error is avoided in two ways. First, large amounts of ammonium salts that remain in solution after precipitation of Cation Group 3 are removed by evaporation and baking; for example, from the skeleton

$$NH_4^{+} + NO_3^{-} \rightarrow N_2(g) + NO_2(g)$$

you can derive an equation by the ion–electron method for the action of nitric acid on ammonium ion. Solid ammonium salts are also decomposed by heating; for example,

$$NH_4Cl(s) \overset{\Delta}{\leftrightharpoons} NH_3(g) + HCl(g)$$

$$NH_4NO_3(s) \overset{\Delta}{\rightarrow} N_2O(g) + 2H_2O(g)$$

A controlled amount of ammonium chloride is then added. Second, the presence of a high concentration of ammonia partially counteracts the effect of ammonium ions. If we put strong players on both teams [numerator and denominator of equation (10.1)], neither can go too far.

In the presence of ammonium ion, the basic ionization of ammonia is sufficiently low that magnesium hydroxide cannot also precipitate with the carbonates of Group 4 (Sec. 2.7).

10.4   Separation and Identification of Barium

Barium chromate is so much less soluble than the chromates of strontium and calcium (Fig. 10.1) that it can be easily separated from the others by control of the chromate ion concentration (Sec. 2.9). Two equilibria are involved:

$$CrO_4^{--} + H^{+} \leftrightharpoons HCrO_4^{-}$$

$$2CrO_4^{--} + 2H^{+} \leftrightharpoons Cr_2O_7^{--} + H_2O$$

Calculations show (Sec. 9.13) that at pH 4.6 appropriate to the ammonium acetate–acetic acid buffer used in the separation, about 72% of the chromium occurs as $Cr_2O_7^{--}$, 27% as $HCrO_4^{-}$, and only 1% as CrO_4^{--}, making the concentration of chromate 6.7×10^{-4} M. This is low enough that strontium chromate will not precipitate but barium chromate will:

$$[Sr^{++}][CrO_4^{--}] < K_{sp}(SrCrO_4) = 2.2 \times 10^{-5} \; M^2$$

$$[Ba^{++}][CrO_4^{--}] > K_{sp}(BaCrO_4) = 1.2 \times 10^{-10} \; M^2$$

The buffer mixture has the advantage of maintaining an almost fixed level of $[H^+]$ and therefore also of $[CrO_4^{--}]$ even though chromate ion is being removed to form $BaCrO_4$.

A strong acid such as HCl will dissolve barium chromate, because it reduces the chromate ion concentration to a very low value. The group precipitate of the carbonates was dissolved in acetic acid, a weak acid, with a glance ahead at the chromate separation, where the hydrogen ion concentration must not be too high. Barium chromate, once it has formed, can be dissolved in a strong acid:

$$2BaCrO_4(s) + 2H^+ \rightarrow 2Ba^{++} + Cr_2O_7^{--} + H_2O$$

Confirmation of barium is obtained by precipitating barium sulfate from this solution. Strontium also forms an insoluble sulfate, but if the conditions of pre-cipitation of $BaCrO_4$ were properly established, little or no Sr^{++} ion should be present. To be on the safe side, we would like to control the sulfate ion con-centration to keep $SrSO_4$ from precipitating. Sulfate is a very weak base, so that its concentration cannot be effectively controlled by the equilibrium

$$SO_4^{--} + H^+ \leftrightharpoons HSO_4^-$$

Instead, we make use of the slow rate of hydrolysis of sulfamic acid in hot acid solution (Sec. 2.3):

$$H_2O + NH_2SO_3^- \xrightarrow{\;H^+\;} SO_4^{--} + NH_4^+$$

The concentration of sulfate ion builds up so slowly that in the course of a few minutes only barium sulfate can precipitate.

10.5   Separation and Identification of Calcium and Strontium

Chromate ion left in the solution after removal of barium chromate must be removed before separating strontium from calcium. This is accomplished by precipitating the carbonates of these cations. A good source of carbonate, such as Na_2CO_3, is satisfactory here, since we are not separating strontium and calcium from other cations but from chromate:

$$Ca^{++} + CO_3^{--} \rightarrow CaCO_3(s)$$

$$Sr^{++} + CO_3^{--} \rightarrow SrCO_3(s)$$

Carbonate ion is then eliminated by treating the precipitate with nitric acid; you should be able to complete

$$CaCO_3(s) + 2H^+ \rightarrow \ ?$$

Strontium can be separated from calcium by making use of the remarkably low solubility of strontium nitrate in 70–72% HNO_3 (the concentrated acid):

$$Sr^{++} + 2NO_3^- \rightarrow Sr(NO_3)_2(s)$$

Other nitrates and chlorides are insoluble in the concentrated acid, $Pb(NO_3)_2$ in HNO_3 and NaCl in HCl, but calcium nitrate is soluble under the conditions of the separation.

After the solid nitrate is dissolved in water,

$$Sr(NO_3)_2(s) \rightarrow Sr^{++}(aq) + 2NO_3^-(aq)$$

strontium ion can be identified by precipitation of strontium sulfate. In the absence of barium, any soluble sulfate such as $(NH_4)_2SO_4$ can be used as a source of sulfate ions. Traces of calcium that may be present can be sequestered (complexed) by triethanolamine

$$Ca^{++} + 2N(C_2H_4OH)_3 \rightarrow Ca[N(C_2H_4OH)_3]_2^{++}$$

This reduces the calcium ion concentration to such an extent that calcium sulfate, which has a comparatively high solubility product, $2.4 \times 10^{-5} \ M^2$, cannot precipitate

$$[Ca^{++}][SO_4^{--}] < K_{sp}$$

The strontium ion, which is larger than Ca^{++} ion, has less tendency to form complexes.

Calcium ion in the solution from the nitric acid separation is precipitated as the oxalate, $CaC_2O_4 \cdot H_2O$, from a hot, weakly alkaline solution. Precipitation is done from hot solution to increase the particle size of the precipitate, an advantage when the precipitate is small and its bulk is to be determined.

It is critically important to obtain sharp separations in the analysis of this group. If barium is not completely removed as chromate, it will precipitate as the nitrate; if strontium is not removed as the nitrate, it will precipitate as the oxalate.

10.6   Some Numerical Illustrations

Fractional precipitation of the sulfates was examined in Sec. 2.2, and the relation between hydrogen and chromate ion concentrations was worked out in Sec. 9.13.

(a) Ammonium Carbonate Solution

In a 0.30 M solution of ammonium carbonate, the 0.60 mol/liter of N atoms and the 0.30 mol/liter of C atoms must be accounted for:

$$0.60 \ M = [NH_3] + [NH_4^+] \tag{10.2}$$

$$0.30 \ M = [CO_3^{--}] + [HCO_3^-] + [CO_2] \tag{10.3}$$

Moreover, the solution should be electrically neutral:

$$[NH_4^+] + [H^+] = [OH^-] + [HCO_3^-] + 2[CO_3^{--}] \tag{10.4}$$

The concentrations of hydrogen and hydroxide ions will be assumed to be negligible in comparison with the rest, and we shall neglect the concentration of CO_2 (or carbonic acid) formed in the second stage of hydrolysis of carbonate ions. We introduce the constants

$$K_a = \frac{[NH_3][H^+]}{[NH_4^+]} = 5.7 \times 10^{-10} \ M \tag{10.5}$$

$$K_2 = \frac{[H^+][CO_3^{--}]}{[HCO_3^-]} = 4.7 \times 10^{-11} \ M \tag{10.6}$$

These five equations can be combined[1] so as to eliminate all variables except $[H^+]$. We then obtain the quadratic equation

$$[H^+]^2 - K_a[H^+] - 2K_aK_2 = 0 \tag{10.7}$$

The solution of this according to the quadratic formula is

$$[H^+] = K_a/2 + [(K_a/2)^2 + 2K_2K_a]^{1/2} \tag{10.8}$$

When the values of the constants are inserted, we have

$$[H^+] = (5.7 \times 10^{-10})/2 + (8.1 \times 10^{-20} + 5.4 \times 10^{-20})^{1/2}$$

whence $[H^+] = 6.5 \times 10^{-10} \ M$. The concentration of carbonate ion can now be found using K_2 and equation (10.3):

$$4.7 \times 10^{-11} = 6.5 \times 10^{-10} \times \frac{[CO_3^{--}]}{0.30 - [CO_3^{--}]}$$

or $[CO_3^{--}] = 2.0 \times 10^{-2} \ M$. This concentration is high enough to precipitate magnesium carbonate if the magnesium concentration is greater than $5 \times 10^{-4} \ M$. The reagent for precipitation of Cation Group 4 is therefore not ammonium carbonate alone but also includes ammonium chloride and ammonia.

Commentary A more accurate calculation would have to take into account the $[CO_2]$ in equation (10.3) and use values of the equilibrium quotients appropriate to the high temperature and ionic strength of the solution. Moreover, ammonium carbonate solutions always contain considerable quantities of ammonium carbamate, NH_4CONH_2. Thus the precise numerical answers of our calculations cannot be taken without a grain of salt, but they indicate the correct order of magnitude of the carbonate ion concentration.

(b) Solubility of Calcium Oxalate and pH

The oxalate ion of calcium oxalate is a basic anion, and the solubility of the precipitate is therefore increased by addition of acid. The relevant equilibria and constants are:

$$CaC_2O_4 \cdot H_2O(s) \rightleftharpoons Ca^{++} + C_2O_4^{--} + H_2O \tag{10.9}$$

$$K_{sp} = [Ca^{++}][C_2O_4^{--}] = 2.1 \times 10^{-9} \ M^2$$

[1] For example, use K_a in (10.2) to eliminate $[NH_3]$ and K_2 in (10.3) to eliminate $[HCO_3^-]$. Then combine with (10.4) and rearrange to give equation (10.7). Try it!

$$H_2C_2O_4 \leftrightharpoons H^+ + HC_2O_4^- \qquad K_1 = \frac{[H^+][HC_2O_4^-]}{[H_2C_2O_4]}$$

$$= 5.4 \times 10^{-2}\ M \qquad (10.10)$$

$$HC_2O_4^- \leftrightharpoons H^+ + C_2O_4^{--} \qquad K_2 = \frac{[H^+][C_2O_4^{--}]}{[HC_2O_4^-]}$$

$$= 5.4 \times 10^{-5}\ M \qquad (10.11)$$

The molar solubility S is equal to the calcium ion concentration if Ca^{++} is the only form of calcium in solution. The oxalate that dissolves distributes itself among the three forms:

$$S = [C_2O_4^{--}] + [HC_2O_4^-] + [H_2C_2O_4] \qquad (10.12)$$

We need only $[C_2O_4^{--}]$ of the three oxalate species for use in K_{sp}, so we eliminate the other two using equations (10.10) and (10.11):

$$S = [C_2O_4^{--}] + \frac{[H^+][C_2O_4^{--}]}{K_2} + \frac{[H^+]^2[C_2O_4^{--}]}{K_1 K_2}$$

$$= [C_2O_4^{--}]\left[1 + \frac{[H^+]}{K_2} + \frac{[H^+]^2}{K_1 K_2}\right] \qquad (10.13)$$

Solving this for the concentration of oxalate, and introducing it and $S = [Ca^{++}]$ into the solubility product expression, we obtain

$$K_{sp} = S \times \frac{S}{1 + ([H^+]/K_2) + ([H^+]^2/K_1 K_2)} \qquad (10.14)$$

or

$$S = \left\{ K_{sp}\left[1 + \frac{[H^+]}{K_2} + \frac{[H^+]^2}{K_1 K_2}\right]\right\}^{1/2} \qquad (10.15)$$

The results of the computation for various pH values are given in Table 10.2. It will be seen that in neutral or basic solutions ($pH \geqslant 7$), the basic nature of the oxalate has a negligible effect and the simple expression $K_{sp} = S^2$ holds.

Table 10.2

Computation of the Solubility of Calcium Oxalate

pH	$\dfrac{[H^+]}{K_2}$	$\dfrac{[H^+]^2}{K_1 K_2}$	$1 + \dfrac{[H^+]}{K_1} + \dfrac{[H^+]^2}{K_1 K_2}$	S, M
1.0	1.85×10^3	3.45×10^3	5300	3.2×10^{-3}
3.0	1.85×10	3.45×10^{-1}	19.8	2.0×10^{-4}
5.0	1.85×10^{-1}	3.45×10^{-5}	1.185	5.0×10^{-5}
7.0	1.85×10^{-3}	3.45×10^{-9}	1.002	4.6×10^{-5}
9.0	1.85×10^{-5}	3.45×10^{-13}	1.000	4.6×10^{-5}

Commentary   The conditions used in the oxalate test in Outline 6 ensure that the solubility will be even lower than that calculated because an excess of oxalate is used. In more precise calculations we should have to take into account the formation of $Ca^{++}C_2O_4^{--}$ ion pairs and the variation in the equilibrium constants with ionic strength. The measured solubility, 5.5×10^{-5} M at pH 6.5, is somewhat higher than we have calculated.

The solubility of carbonates as a function of pH can be treated in a similar fashion. Carbonate ion is a much stronger base than oxalate. The deposition of calcium carbonate in caves or sea water is an interesting geochemical problem.

EXPERIMENTAL PART

10.7   Analysis of Known and Unknown Samples

(1) A *known* or *practice solution* for this group should contain 3 mg/ml of Ba^{++}, Sr^{++}, Ca^{++}, and Mg^{++} ions and about 30 mg/ml of NH_4^+ ion. Use 10 drops for the analysis. Remove the ammonium salts and analyze the group according to Outline 6. As a check on your technique, add ammonium sulfate and ammonium oxalate to Solution 6.1 as directed in Sec. 11.9(b)(3). There should be no more than a trace of precipitate even after warming. Add Na_2HPO_4 solution to this to precipitate magnesium ammonium phosphate and verify that magnesium was not precipitated with Group 4.

(2) The *unknown* may contain the ions of Group 4 alone or of Groups 4 and 5 together. If it is a solution, use 10 drops for the analysis according to Outline 6. If it is a solid, take a representative sample (T-1) and dissolve 20–30 mg in water or 6 M HCl and use this solution for the analysis. Remove ammonium salts only if a positive test for the ammonium ion is obtained (Preliminary Test, Outline 7).

If the unknown is Solution 5.1 from Cation Group 3, be sure it has been evaporated with hydrochloric acid as directed in Outline 5 to remove thioacetamide and hydrogen sulfide before any nitric acid is added, for the latter can oxidize sulfide to sulfate and cause premature precipitation of barium and strontium.

10.8   Flame Tests

Characteristic flame colors are given by all three elements of this group: orange red for calcium, carmine for strontium, and yellowish green for barium. When observed through a spectroscope, the flame colors are resolved into bands of color (Fig. 10.2) that are really groups of closely spaced lines and are given by molecules, in this case metal oxides formed by oxidation of the metal salts in the flame. The alkali metals give single sharp lines originating in excitation of their atoms.

For practice, place 1 or 2 drops of each of the standard solutions of the ions of this group in separate test tubes, and to each add 2 drops of 12 M HCl (metal chlorides are more volatile than other salts such as nitrates). Clean a Nichrome (or platinum) wire by heating it for some time in the flame; if necessary, dip the hot wire in 12 M HCl to dissolve residual salts; persistent traces can be disposed of by snipping off the tip of the Nichrome with scissors (*don't* snip platinum). Then test the flame color of each ion. Clean the wire between tests. If a satisfactory flame test is not obtained from a solution, place a small portion of the solid chloride (side shelf) in a depression of a spot plate, moisten it with 12 M HCl, and dip the wire into this sample. Note the color and persistence of each flame.

No flame color is altogether satisfactory by itself for identification of the ion. Other elements give flame colors that can obscure or be confused with those of the alkaline earth metals; for example, copper can be confused with barium, and sodium can obscure barium. Furthermore, the flame color of any one of the Group 4 elements is obscured by those of the other two when the latter are present in large excess. Thus, in the hands of the inexperienced, these flame tests are not as reliable as the precipitation tests, but they are useful supplements to the latter.

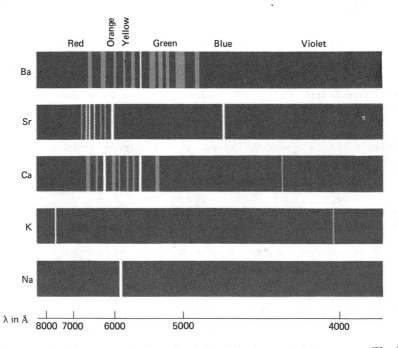

Fig. 10.2. Flame spectra. White lines stand for the most intense ones. The fine structure of the bands is not shown.

OUTLINE 6. The Systematic Analysis of Cation Group 4

Removal of excess ammonium salts. To soln under test in a casserole, add 5 drops 16 M HNO$_3$ and evap to dryness [T-7(b)] over a microflame. Bake walls and bottom of the dish in the flame until dense white clouds of ammonium salts are no longer evolved.[1] Keep temperature below red heat.[2] Cool. Add several drops 6 M HCl[3] and with a rod rub the acid against the wall and bottom of the casserole to help bring residue into soln.[4] Dilute with about 10 drops water and transfer soln to a centrifuge tube. Rinse casserole with another 5–10 drops water and add the rinsing to the first soln. If soln is not clear, centrifuge and discard the residue.[5]

Precipitation of the group. To the soln in the centrifuge tube, add 1 drop 6 M NH$_4$Cl. Then make alk with 15 M NH$_3$ (T-12).[6] Add 5 drops (NH$_4$)$_2$CO$_3$ soln and stir. Set the tube in a bath of *warm*, not hot, water only long enough to coagulate the ppt (not more than 1 minute).[7] Centrifuge and separate Soln 6.1. Wash the pptd carbonates with a little water.[8]

Solution 6.1. Mg^{++}, K$^+$, Na$^+$, NH$_4$ salts, NH$_3$. Make sure that soln is alk. Then add 1 drop (NH$_4$)$_2$CO$_3$. If more ppt forms, combine it with Ppt 6.1. If pptn is complete, acidify soln with HCl and set aside for the analysis of **Group 5.**

Precipitate 6.1. BaCO$_3$, SrCO$_3$, CaCO$_3$.[9] Dissolve in a min of 6 M HOAc. Any residue insol in the acid can be removed and discarded. To the soln (Ba^{++}, Sr^{++}, Ca^{++}) add several drops 3 M NH$_4$OAc and dil with 10 drops water. Add a few drops K$_2$CrO$_4$ soln. Centrifuge. Wash ppt with water.

Precipitate 6.2. Yellow BaCrO$_4$, (SrCrO$_4$). Dissolve in 1–2 drops 6 M HCl. Dilute with about 1 ml water. Add one-fourth spatula-full HNH$_2$SO$_3$ and stir to dissolve. Heat in a bath of vigorously boiling water for 10 minutes.

White cryst ppt, BaSO$_4$: **presence of Ba.**

Solution 6.2. Sr^{++}, Ca^{++}, CrO$_4^{--}$. If soln is not bright yellow add more K$_2$CrO$_4$. Make soln alk with 6 M NaOH[10] and add 5 drops freshly prepared Na$_2$CO$_3$ soln.[11] Warm to coagulate ppt. (If no ppt forms, Sr and Ca are absent.) Centrifuge and withdraw soln to a centrifuge tube and add more Na$_2$CO$_3$ soln. If no ppt forms on warming, discard soln. Wash the pptd carbonates twice with water to remove CrO$_4^{--}$.

Precipitate 6.3. SrCO$_3$, CaCO$_3$. Dissolve in a few drops HNO$_3$ and transfer to a casserole. Rinse tube with water and add rinsings to casserole. Evap soln over a microflame almost to dryness; withdraw from flame and let heat of dish complete the evapn. Cool. Add 5 drops 16 M HNO$_3$. Rub inside of casserole with a *dry* stirring rod to loosen ppt; and pour mixture quickly[12] into a *dry* centrifuge tube.[13] Rinse the casserole with 5 more drops 16 M HNO$_3$ and pour into the tube. Insert a *dry* rod and stir and rub the wall. Let stand for at least 5 minutes in cold water with occasional stirring.[14] Centrifuge and wash ppt with 16 M HNO$_3$.

Precipitate 6.4
(see facing page)

Solution 6.4
(see facing page)

┌──────────── **Precipitate 6.3** ────────────┐
↓ ↓

Precipitate 6.4. Sr(NO₃)₂, [Ca(NO₃)₂]. Dissolve in 5–10 drops water. Make alk with NH_3 and add several drops of triethanolamine and a few drops $(NH_4)_2SO_4$ soln. Rub the inner wall with a rod and warm.[15] White crystn ppt, $SrSO_4$: **presence of Sr.**[16]

Solution 6.4. Ca^{++}, HNO_3. Dilute with about 1 ml water. Make slightly alk with 15 M NH_3.[17] Add a few drops $(NH_4)_2C_2O_4$ soln. Smoky white ppt, CaC_2O_4: **presence of Ca.**

Notes on Outline 6

1. Use a somewhat higher flame in the baking than in the evaporation and play it over the walls as well as the bottom of the casserole. Otherwise, ammonium salts after being vaporized may re-form on the cooler upper walls.

2. If too high a temperature is reached, alkali halides may become fused into the glaze of the dish and lost.

3. Magnesium chloride may lose HCl at high temperatures during removal of the water. The basic chloride that is formed, $Mg_2(OH)_3Cl \cdot 4H_2O(s)$, is not soluble in water but dissolves in HCl.

4. The baked residue goes into solution slowly and adheres to the casserole.

5. In a general analysis, Solution 5.1 will contain organic matter from the hydrolysis of thioacetamide. Some of this is destroyed by the nitric acid; the rest is charred during baking. This accounts for a black residue of carbon often obtained at this point.

6. Recall that mixing is difficult in a centrifuge tube because of the tapered bottom. Work the glass rod up and down to prevent layering of the NH_3 solution on top of the solution under test (T-5). If you fail to get a group precipitate, the solution may not be basic.

7. Ammonium carbonate decomposes completely into NH_3, CO_2, and H_2O if heating is prolonged, and precipitation of the carbonates would then be incomplete.

8. In washing these dense precipitates be careful to disperse them completely in the wash water.

9. The precipitates of this group are more densely packed than the sulfides and hydroxides of preceding groups, but although they appear small, it is possible to obtain a satisfactory test with them.

10. Disregard the formation of any precipitate after addition of NaOH. Sodium hydroxide solution is frequently contaminated with carbonate, so it may be the carbonates of strontium and calcium or it may be magnesium hydroxide if some magnesium carbonate came down with Group 4.

11. Dissolve a heaping spatula-full (about 0.15 g) of solid Na_2CO_3 in 10 drops of water. When such solutions are stored in glass bottles, they etch the glass and dissolve some silica. This would precipitate with the carbonates of strontium and calcium and make it difficult to judge the quantity of these substances. It is best to use a fresh solution of the reagent.

12. The solid nitrate settles rapidly. Unless the mixture is poured quickly, only the supernatant solution is decanted; the precipitate is left behind in the casserole. Examine the casserole if no precipitate is obtained to see if any crystals are adhering to the walls.

13. For a satisfactory separation of $Sr(NO_3)_2$, the concentration of HNO_3 must not be below 70% (16 M); ideally, it should be close to 80%. Thus the introduction of water

must be avoided by (1) using a dry tube, (2) pouring instead of pipetting, and (3) using a dry rod. Dry the tube with a *clean* towel or in an oven. CAUTION: 16 M HNO$_3$ is very corrosive. It will produce yellow stains on skin that take a week to wear (or peel) off. Flood with water all spills of nitric acid, whether on the skin or on the desk. The use of plastic gloves is recommended.

14. Strontium nitrate tends to supersaturate. Stirring, rubbing, and standing are the usual techniques of coping with supersaturation. Cooling also helps, not because the solubility is appreciably decreased, but because the supersaturated solution becomes more unstable.

15. Strontium sulfate also tends to supersaturate. Allow at least 5 minutes for it to form.

16. If desired, a flame test can be tried for confirmation. The sulfate must be transposed to carbonate and dissolved in hydrochloric acid. Remove the solution above the precipitate, and treat the latter with ammonium carbonate solution. Stir up the cake and warm it briefly. Centrifuge and wash the precipitate several times to remove sulfate. Dissolve in a few drops of 12 M HCl and do the flame test (Sec. 10.8).

17. Since 10 drops of 16 M HNO$_3$ were used in the separation of Sr(NO$_3$)$_2$, it will take at least this much 15 M NH$_3$ for the neutralization. The nitric acid was diluted with water to make the reaction less violent. Failure to make the solution basic is a common error at this stage. Use particular care in mixing (T-5) and testing with litmus (T-12).

EXERCISES

10.1. Name the ions of this group, tell why they occur together, and place them in the periodic table.

10.2. What are some consequences of the smaller size and higher charge of Ca^{++} as compared with K$^+$?

10.3. Account for the fact that ions of this group form soluble chlorides but insoluble fluorides, and soluble nitrates but insoluble carbonates.

10.4. Write an equation for the proton transfer reaction between ammonium and carbonate ions. What is the effect on this equilibrium of (a) addition of ammonium nitrate, (b) addition of ammonia, (c) heat?

10.5. Explain why a combination of ammonium chloride and ammonia is used along with ammonium carbonate to precipitate the group.

10.6. Explain why barium is separated as the chromate rather than as the sulfate although the latter is more insoluble.

10.7. Explain how the concentration of chromate ion can be controlled by a mixture of ammonium acetate and acetic acid.

10.8. In the precipitation of barium chromate, what harm would result if acetic acid were replaced by (a) hydrochloric acid, (b) ammonia?

10.9. Explain each of the following in the systematic analysis:
(a) Triethanolamine is present during the precipitation of SrSO$_4$.
(b) Ammonium salts are removed before precipitation of the group.
(c) Dry equipment is used in precipitating Sr(NO$_3$)$_2$.
(d) Barium sulfate is precipitated with sulfamic acid but strontium sulfate is precipitated with ammonium sulfate.

(e) A freshly prepared solution of sodium carbonate is used to precipitate the carbonates of Sr and Ca.

(f) The solution is heated when $BaSO_4$ is precipitated.

10.10. What difficulties would result if

(a) The group precipitate were dissolved in HCl.

(b) Ammonium salts were not removed before precipitation of the group.

(c) Sodium carbonate were used to precipitate the group.

(d) Ammonia were not used in precipitation of the group.

10.11. Write equations for the successive reactions of each ion of this group from precipitation to identification.

10.12. What procedure would you follow to answer each of the following questions

(a) Is a white pigment lithopone $(ZnS + BaSO_4)$ or white lead $[Pb_3(OH)_2(CO_3)_2]$?

(b) Is the acid constituent of a baking powder a phosphate $[Ca(H_2PO_4)_2]$, or alum $[NaAl(SO_4)_2 \cdot 12H_2O]$, or is it "double acting" (containing both)? (The powder will also contain $NaHCO_3$ and perhaps other constituents such as calcium sulfate and starch.)

(c) Is a white powder barium meal ($BaSO_4$, used to facilitate X-ray pictures of the stomach and intestines) or plaster of Paris ($CaSO_4 \cdot \frac{1}{2}H_2O$)?

10.13. The rate constant for hydrolysis of sulfamic acid (Sec. 2.3) is 1.3 liter mol^{-1} sec^{-1} at 80.35°C, 4.2 at 90°C, and 10.5 at 98°C.

(a) What would be the approximate relative rate of hydrolysis of equimolar solutions of sulfamic acid at 80.35°C and 98°C?

(b) Which would hydrolyze faster: a mixture at 90°C that is 0.10 M in H$^+$ and $NH_2SO_3^-$ ions or a mixture at 98°C that is 0.050 M in each ion?

10.14. Calculate the chromate ion concentration in a solution that has a pH of 1.0 and is 0.050 M in total chromium. If the solution is also 0.0010 M in barium ion, will barium chromate precipitate? (See Sec. 9.13.)

10.15. It is proposed to separate calcium and strontium ions by fractional precipitation of their oxalates. An organic ester of oxalic acid is slowly hydrolyzed, giving rise to a gradual increase in oxalate ion concentration. Suppose the solution is initially 0.020 M in Ca^{++} and 0.0010 M in Sr^{++}.

(a) Which will precipitate first, CaC_2O_4 or SrC_2O_4?

(b) What will be the concentration of oxalate ion when the second precipitate begins to form?

(c) What percentage of the first ion to come down as the oxalate is left in solution when the second oxalate begins to precipitate?

(d) Is this a practicable method of separating strontium from calcium?

10.16. Calculate the molar solubility of $BaSO_4$ at various pH values between 0 and 5 and compare the effect of pH on the solubilities of $BaSO_4$ and $CaC_2O_4 \cdot H_2O$ (Sec. 10.6). Assume the reciprocal of $K_a(H_2SO_4)$ to be zero.

ELEVEN

Cation Group 5. The Soluble Group: $Na^+, K^+, Mg^{++}, NH_4^+$

11.1 Introduction

The ions in this group are left in solution after separation of all the other cations. The position in the periodic table of the three metals and of less common members of the Soluble Group are shown below.

IA	IIA
(Li^I)	
Na^I	Mg^{II}
K^I	
(Rb^I)	
(Cs^I)	
(Fr^I)	

11.2   General Characteristics of Sodium and Potassium Ions

Sodium and potassium are alkali metals, members of periodic Group IA. Their atoms have a single s valence electron, which is lost to give a univalent ion with electron configuration of a noble gas. Such d^0 ions—so-called because their outer shells contain no d electrons—give colorless compounds unless the anion is colored, as it is in yellow Na_2CrO_4 and purple $KMnO_4$. Color is associated with absorption of light energy to change the electronic state of the ion. Visible light does not carry sufficient energy to disturb the stable electronic configuration of the alkali metal ions and is not affected by them.

Sodium and potassium ions are the largest cations in their periods: radii 0.98 Å for Na^+, 1.33 Å for K^+. Their comparatively large size, low charge, and d^0 (noble gas) structure allow them to exert only weak attraction for other ions or molecules. In aqueous solution they attract the oxygen ends of water molecules and are hydrated. Solid sodium salts are frequently crystallized with water of hydration; for example, $NaI \cdot 2H_2O$, $Na_2SO_4 \cdot 10H_2O$. Aside from the hydrated ions, sodium and potassium form very few complexes. The recently discovered "crown" complexes depend for their stability on the fit between the hydrated Na^+ or K^+ ion and the hole in the center of the ligand (see diagram).

Dicyclohexyl-16-crown-5 Na^+ complex

Almost all compounds of the two ions are comparatively soluble in water. The attraction between these large, low-charged ions and anions in the crystals is not sufficiently strong to resist disruption of the crystals by water. For many salts, particularly those of weak acids, the solubilities of the potassium compounds are greater than those of sodium, an order attributable to the decrease in stability or lattice energy of the crystals as the size of the cation increases. Crystals containing bulky anions such as ClO_4^-, $PtCl_6^{--}$, or $Co(NO_2)_6^{3-}$, are all comparatively weakly put together; the sodium salts are then the more soluble, because hydration stabilizes Na^+ in solution more than K^+.

11.3   Detection of Sodium

A triple acetate of sodium is sufficiently insoluble to be precipitated from moderately dilute solutions:

$$Na^+ + Mg^{++} + 3UO_2^{++} + 9OAc^- + 9H_2O \rightarrow NaMg(UO_2)_3(OAc)_9 \cdot 9H_2O(s)$$

It is a bulky precipitate, an advantage when only a small amount of sodium has to be detected: 100 mg of the precipitate is obtained from only 1.53 mg of sodium.

The precipitate is fairly soluble, and careful control of conditions is required. Even at best, precipitation will occur only when the concentration of sodium exceeds 0.7 mg/ml. The loss of sodium because of the solubility of the precipitate can be decreased by keeping low the volume of solution under test and adding an excess of reagent. The latter is a mixture of magnesium acetate, uranyl acetate, and acetic acid saturated with the sodium compound. The ions that interfere with this test are (1) H^+ from strong acid, for it consumes acetate to give molecular acetic acid; (2) K^+ if present at a concentration much larger than 50 mg/ml, for it precipitates a pale yellow compound; (3) Li^+, Ag^+, Hg^{++}, and Sr^{++}, which also form precipitates; (4) Bi^{3+}, Sb(III), and Sn(IV), which hydrolyze to give precipitates in the weakly acidic solution; and (5) PO_4^{3-}, AsO_4^{3-}, OH^-, and other anions that form precipitates with uranyl ion.

Sodium gives a characteristic persistent yellow flame test caused by emission of light of wavelengths 5890 and 5896 Å. This test is too sensitive to use by itself— even traces of sodium give a strong yellow color to the flame.

11.4   Detection of Potassium

There are a number of slightly soluble potassium salts with bulky anions; we use the cobaltinitrite[1] salt

$$2K^+ + Na^+ + Co(NO_2)_6^{3-} + H_2O \rightarrow K_2NaCo(NO_2)_6 \cdot H_2O(s)$$

Its saturated solution is about 0.001 M, and as little as 0.02 mg of K^+ per ml can be detected. The composition of the precipitate varies somewhat with temperature and concentration. There are a number of interfering ions: (1) NH_4^+ forms a

---

[1] The systematic name of $Co(NO_2)_6^{3-}$ is hexanitrocobaltate(III) ion. In the familiar name cobalt*initr*ite note the i–i–i.

similar precipitate, $(NH_4)_2NaCo(NO_2)_6 \cdot H_2O$; (2) H^+ from strong acid decomposes nitrite

$$3NO_2^- + 2H^+ \rightarrow 2NO(g) + NO_3^- + H_2O$$

and Co(III) is reduced to pink Co^{++}; (3) OH^- from strong base precipitates black $Co(OH)_3$; (4) oxidizing or reducing agents destroy the reagent; and (5) readily hydrolyzable ions give precipitates in the weakly acid solution.

Potassium gives a brief violet flame color, easily masked by sodium, caused by emission of light of wavelengths 4045 Å (violet) and 7665 and 7699 Å (red). Cobalt glass can be used to block the yellow sodium light and permit the potassium flame color to be observed. The variation of transmittance of cobalt glass with wavelength of light is shown in Fig. 11.1. The glass transmits light only at the ends of the visible spectrum, and these are the very regions in which the spectral lines of potassium lie.

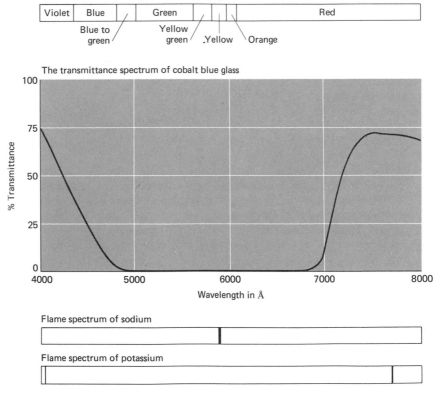

Fig. 11.1. The transmittance spectrum of cobalt blue glass. The glass is opaque to light with wavelengths between 5000 and 6700 Å. At 4000 Å two thicknesses of blue glass transmit 75% as much light as the same thickness of clear glass. The colors of the visible spectrum and the lines in the flame spectra of sodium and potassium are given on the same wavelength scale.

11.5   Some Properties of Magnesium Ion

Magnesium atoms have a pair of $3s$ valence electrons that are lost to give the d^0 ion Mg^{++}. Like other d^0 ions, Mg^{++} has a stable electronic configuration and its compounds are colorless unless the anion is colored. Magnesium and sodium ions have the same electronic configuration, but because of the higher nuclear charge of magnesium, the radius of its ion is about the same as that of Li^+, namely, 0.65 Å.

Magnesium ion, because of its large ratio of charge to radius, forms many more complexes than sodium or potassium ions do. Most of these have Mg—O bonds, but none is particularly stable in solution or useful in qualitative analysis. The chlorophylls are notable examples of compounds in which Mg^{++} ion is co-ordinated to nitrogen.

All magnesium salts, with the possible exception of magnesium oxalate, are highly ionized. Because of the strong attraction of magnesium ion for water, many magnesium salts are highly hydrated; for example, $MgCl_2 \cdot 6H_2O$, $MgSO_4 \cdot 7H_2O$ (epsom salt), and $Mg(NO_3)_2 \cdot 6H_2O$. Hydration promotes solubility by stabilizing the ion in solution; the chloride, bromide, iodide, sulfate, nitrate, and chromate are all soluble salts, and its carbonate is more soluble than that of the less hydrated calcium ion. Magnesium fluoride, in which there are strong bonds in the crystal between Mg^{++} ion and the small F^- ion, is moderately insoluble. The most common slightly soluble compounds are the hydroxide, basic carbonate, and several phosphates. In the presence of ammonia and ammonium chloride, the addition of Na_2HPO_4 to solutions of magnesium salts causes precipitation of magnesium ammonium phosphate:

$$Mg^{++} + NH_4^+ + PO_4^{3-} + 6H_2O \rightarrow MgNH_4PO_4 \cdot 6H_2O(s)$$

This is the best reaction for separation of magnesium from other ions, because the precipitate is crystalline and hence purer than gelatinous compounds such as $Mg(OH)_2$ and $MgHPO_4$. It is also the least soluble of the common salts of magnesium.

11.6   Separation and Detection of Magnesium

In order to get a successful separation of magnesium as magnesium ammonium phosphate, several conditions must be established.

(1) Other ions precipitated by hydroxide or phosphate must be absent. Traces of Cation Group 4 that escaped precipitation as carbonates are removed by addition of ammonium oxalate and ammonium sulfate:

$$Ca^{++} + C_2O_4^{--} \rightarrow CaC_2O_4(s)$$
$$Sr^{++} + SO_4^{--} \rightarrow SrSO_4(s)$$
$$Ba^{++} + SO_4^{--} \rightarrow BaSO_4(s)$$

(2) A combination of ammonia and ammonium salt controls the hydroxide ion concentration (Sec. 2.7) so that magnesium hydroxide cannot precipitate but a high enough concentration of phosphate is present (Sec. 11.8).

(3) Supersaturation is overcome by stirring and rubbing the inner wall of the vessel.

The presence of magnesium in the precipitate can be confirmed by dissolving it in acid

$$MgNH_4PO_4 \cdot 6H_2O(s) + 2H^+ \rightarrow Mg^{++} + NH_4^+ + H_2PO_4^- + 6H_2O$$

and then forming a colored "lake,"[2] a precipitate of magnesium hydroxide colored by an adsorbed dye. S. and O. reagent,[3] $(HO)_2C_6H_3N{=}NC_6H_4NO_2$, gives a blue lake. As little as 0.002 mg of magnesium can be detected if other elements that form insoluble hydroxides and large amounts of ammonium salts are absent.

11.7 Some Properties of Ammonium Ion and Its Detection

The NH_4^+ ion is roughly spherical with a radius (1.43 Å) close to that of potassium (1.33 Å). Its salts resemble those of potassium in solubility; for example,

$$2NH_4^+ + Na^+ + Co(NO_2)_6^{3-} + 6H_2O \rightarrow (NH_4)_2NaCo(NO_2)_6 \cdot 6H_2O(s)$$

Ammonium ion must be removed before the detection of potassium. Fortunately, it differs from potassium in several ways that make it possible to identify or remove it.

Unlike potassium ion, ammonium ion is a proton donor, and its conjugate base is the volatile gas ammonia. The ion is therefore detected by converting it to ammonia with an excess of strong base:

$$NH_4^+ + OH^- \rightarrow NH_3(g) + H_2O$$

The liberated ammonia is most easily detected by absorbing it in moist red litmus paper. The above reaction reverses to a sufficient extent to turn the litmus blue.

Destruction of ammonium salts is possible by either thermal decomposition or oxidation. When thermal decomposition involves only proton transfer, it is reversible:

$$NH_4Cl(s) \overset{\Delta}{\rightleftharpoons} NH_3(g) + HCl(g)$$

$$(NH_4)_2SO_4(s) \overset{\Delta}{\rightleftharpoons} 2NH_3(g) + SO_3(g) + H_2O(g)$$

---

[2] This term has no connection with bodies of water but is a variant of the word "lac," which is applied to a resin that is the source of a crimson dye.

[3] Named after Suitzu and Okuma, who investigated its behavior. Its scientific name is *para*-nitrobenzeneazoresorcinol (accent -ben- and -sor- and use it to frighten children, astonish parents, and bore friends).

Oxidation is a more efficient way of destroying ammonium ion because it is not reversible:

$$NH_4NO_3(s) \xrightarrow{\Delta} N_2O(g) + 2H_2O(g)$$

$$NH_4^+(aq) + NO_2^-(aq) \rightarrow N_2(g) + 2H_2O$$

$$4H^+ + 6NO_3^- + 2NH_4^+ \rightarrow N_2(g) + 6NO_2 + 6H_2O$$

The last reaction occurs when solutions of ammonium salts are evaporated with concentrated nitric acid.

11.8 A Numerical Example

The failure of magnesium hydroxide to precipitate from ammonia–ammonium chloride buffer was verified by computation in Sec. 2.8.

Precipitation of Magnesium Ammonium Phosphate

During the analysis of the known solution, this precipitate is obtained from a mixture that contains $\frac{1}{3}$ mg of Mg^{++} ion, one drop each of 6 M NH_3 and 6 M NH_4Cl, and 2 drops of 0.5 M Na_2HPO_4 in a total volume of 7 drops ($\frac{1}{3}$ ml). The initial concentrations are therefore $[Mg^{++}] = 0.04$ M; $[NH_3] = [NH_4^+] = 0.9$ M; and total phosphate = 0.14 M. The pertinent equilibria and constants are

$$MgNH_4PO_4 \cdot 6H_2O(s) \leftrightharpoons Mg^{++} + NH_4^+ + PO_4^{3-} + 6H_2O$$

$$K_{sp} = [Mg^{++}][NH_4^+][PO_4^{3-}] = 7.1 \times 10^{-14} \, M^3$$

$$HPO_4^{--} \leftrightharpoons H^+ + PO_4^{3-} \qquad K_3' = 4.2 \times 10^{-13} \, M$$

$$H_2PO_4^- \leftrightharpoons H^+ + HPO_4^- \qquad K_2 = 6.3 \times 10^{-8} \, M$$

and

$$NH_4^+ \leftrightharpoons H^+ + NH_3 \qquad K_a = 5.7 \times 10^{-10} \, M$$

Since $[NH_3] = [NH_4^+]$, we can set $[H^+]$ equal to K_a:

$$K_a = [H^+]\frac{[NH_3]}{[NH_4^+]} = [H^+] = 5.7 \times 10^{-10} \, M$$

The ratios of concentrations of the various phosphate ions can be obtained from K_2 and K_3:

$$\frac{[PO_4^{3-}]}{[HPO_4^{--}]} = \frac{K_3}{[H^+]} = \frac{4.2 \times 10^{-13}}{5.7 \times 10^{-10}} = 7.4 \times 10^{-4}$$

$$\frac{[H_2PO_4^-]}{[HPO_4^{--}]} = \frac{[H^+]}{K_2} = \frac{5.7 \times 10^{-10}}{6.3 \times 10^{-8}} = 9 \times 10^{-3}$$

Clearly, most of the phosphate is in the form of HPO_4^{--} ions, and we can approximate $[HPO_4^{--}]$ by the total phosphate concentration, 0.14 M. It follows that $[PO_4{}^{3-}] = [HPO_4{}^{--}] \times K_3/[H^+] = 0.14 \times 7.4 \times 10^{-4} = 1.0 \times 10^{-4}\ M$. The ion concentration product of magnesium ammonium phosphate is then

$$[Mg^{++}][NH_4{}^+][PO_4{}^{3-}] = 4 \times 10^{-2} \times 0.9 \times 1.0 \times 10^{-4} = 3.6 \times 10^{-6}$$

This is much larger than the equilibrium value, K_{sp}, so we expect the precipitate to form even though most of the phosphate is in the form of $HPO_4{}^{--}$.

Magnesium hydroxide cannot precipitate under these conditions. The hydroxide ion concentration of this mixture is

$$[OH^-] = \frac{K_w}{[H^+]} = K_b(NH_3) = 1.8 \times 10^{-5}$$

and the ion concentration product is

$$[Mg^{++}][OH^-]^2 = (4 \times 10^{-2})(1.8 \times 10^{-5})^2 = 1.3 \times 10^{-11}\ M^3$$

The solubility product of active (freshly precipitated) magnesium hydroxide is about $6.3 \times 10^{-10}\ M^3$. We therefore have the relation $[Mg^{++}][OH^-]^2 < K_{sp}$, and no precipitate can form.

EXPERIMENTAL PART

11.9   Analysis of Known and Unknown Samples

(a) Use 10 drops of a *known* or *practice solution* that contains about 2 mg of Mg^{++}, 2 mg of $NH_4{}^+$, 4 mg of K^+, and 4 mg of Na^+ per ml.

(b) The *unknown* may be

    (1) A solution containing only the ions of Group 5. Use 10 drops in testing for Mg^{++}, K^+, and Na^+ and a separate portion for the test for $NH_4{}^+$.

    (2) A solid mixture containing only the ions of Group 5. Take a representative sample (T-1) and use 20 mg [T-3(b)] for the tests for Mg^{++}, Na^+, and K^+. Dissolve in 8–10 drops of water. If the solution is acid, make it just alkaline with 6 M NH_3 [check with litmus (T-12)]. If the solution is alkaline, acidify with 6 M HCl and then make it barely alkaline with 6 M NH_3.

    (3) Solution 6.1 (Outline 6, Cation Group 4) obtained in a general analysis after removal of Groups 1–4. Use the entire solution for the tests for Mg^{++}, K^+, and Na^+. Add one drop each of $(NH_4)_2C_2O_4$ and $(NH_4)_2SO_4$ solutions and 6 M NH_3, if necessary, to make the solution alkaline. Centrifuge (T-8) to settle the precipitate, which will contain traces of the ions of Group 4 that escaped precipitation. Withdraw the solution and use it for the analysis; discard the residue. The test for the ammonium ion must be done on a separate portion of the *original unknown*.

11.10   Flame Tests

Clean a Nichrome (or platinum) wire by dipping it in a little 12 M HCl and holding it in the flame. One end of the Nichrome wire can be thrust into a cork to make a convenient holder. Repeat the treatment until the wire ceases to color the flame (the wire itself will be red hot). Then touch the wire to the sample under test and bring it back to the flame. If the flame is colored yellow by sodium, the violet color of potassium may be masked. Dip the wire in the sample again and view the flame through two thicknesses of cobalt blue glass. The potassium flame looks red through the glass. Since the test is fleeting, have the glass and your eyes in position before you bring the wire into the flame.

It is suggested that you try the flame tests on known samples, solids or solutions, before attempting the unknown. To convince yourself that the tests are not sure-fire but need interpretation, try the following experiments.

(1) Mix a few crystals of sodium and potassium chlorides in a depression of the spot plate and try the flame test on this.

(2) Mix one drop each of the standard solutions of Na^+ and K^+ with 1 ml of 6 M HCl and try the test on this.

(3) Draw the clean, cool wire between your fingers and bring it into the flame.

OUTLINE 7. The Systematic Analysis of Cation Group 5

Preliminary test for the ammonium ion. This must be done on original sample, for ammonia and ammonium salts are added during the analysis for the other ions.

In a 5-ml beaker or crucible, put 2 drops soln or 10 mg solid sample. Add a small piece of red litmus paper. Cut another piece of red litmus paper smaller than the diameter of the beaker, moisten it, and attach it to the convex side of a watch glass. Add 6 M NaOH, drop by drop, to contents of beaker, stirring after each addition, until soln is alkaline; then add a few[1] drops in xs.[2] (If metal ions of Groups 1–3 are present, then add 1–2 drops Na_2S soln.[3]) Cover immediately with the watch glass so that the litmus paper is centered over the soln.[4] If the paper turns blue within 2 minutes, report **presence of NH_4^+**.[5] Check your technique by carrying out the same test on a sample of distilled water.

Analysis for Mg^{++}, K^+, and Na^+. Start with a known or unknown soln (Sec. 11.9). Divide into two parts: one-third in a centrifuge tube for Mg test, two-thirds in a casserole for K and Na tests.

Mg test. Soln should be alk. Add 1 drop 6 M NH_4Cl if ammonium salts are not already present. Then add a few[1] drops Na_2HPO_4 soln. Stir and rub the inner bottom wall of the tube with a rod. Set aside for at least 5 minutes.[6] Centrifuge (T-8)[7] and draw off and discard the soln.(T-9).

White cryst ppt $MgNH_4PO_4 \cdot 6H_2O$. Dissolve in a few[1] drops 6 M HCl. Add 1 drop (no more) S. and O. reagent and make soln alk with 6 M NaOH (T-12). If soln and ppt are orange, add another drop NaOH. Centrifuge to settle ppt so that it can be observed more readily.[8]

Deep blue lake, $Mg(OH)_2$: **presence of Mg.**

K and Na tests. Removal of NH_4 salts. Evap to dryness [T-7(b)][7] over a microflame [T-7(a)].[9] Cool. Add 10 drops 16 M HNO_3 and manipulate casserole so that acid wets wall as well as bottom. Again evap to dryness either under ventilating duct or in hood. Bake residue until clouds of ammonium salts cease to come off. Heat walls as well as bottom of casserole, but do not heat to redness.[10] Cool almost to room temp. Then add 4 drops H_2O and rub inside of casserole with a rod to loosen caked residue. Transfer to a centrifuge tube and centrifuge if soln is turbid (T-8). Discard residue and divide soln equally between two centrifuge tubes. Avoid diluting it.[11]

Na test. Add 6 drops magnesium uranyl acetate. Stir and rub inside wall of tube. Let stand for 5 minutes[15] with occasional stirring.

Pale, greenish yellow cryst ppt, $NaMg(UO_2)_3(OAc)_9 \cdot 9H_2O$: **presence of Na.**

K test. Add 1 drop 6 M HOAc and a single crystal of $NaNO_2$. Warm to destroy traces of NH_4^+. Cool and add a few drops $Na_3Co(NO_2)_6$ reagent.[12] If a ppt forms, warm for several minutes in the hot water bath.[13] Cool and add more reagent.

Yellow ppt, $K_2NaCo(NO_2)_6 \cdot H_2O$: **presence of K.**

Notes on Outline 7

1. The words "a few" generally mean 1 or 2, and "several" is taken to mean 2 to 4. The directions have purposely been left vague so that you will learn to judge for yourself how much reagent to add. This is better than slavishly adding a set number of drops regardless of the quantity of material under test. It is wise to be conservative in adding reagent, although inexperience may lead some to believe that if one drop is good, a liter is better.

2. Avoid a large excess of base, for this dilutes the solution so that not enough ammonia gas may be liberated to change the color of litmus.

3. Certain cations, such as Cu^{++}, Ag^+, and Zn^{++}, form complex ions with ammonia and hinder its volatilization. The sodium sulfide removes these cations as insoluble sulfides.

4. If the litmus touches the sides of the beaker and they are wet with NaOH, a false test can be obtained.

5. The sensitivity of the test can be increased by (1) concentrating the solution before the test by evaporation and (2) warming the solution before adding the NaOH. These precautions are usually not necessary. The paper may turn blue on prolonged exposure to laboratory air, which frequently contains ammonia fumes.

6. The magnesium ammonium phosphate tends to supersaturate. Stirring and rubbing the walls helps induce crystallization. Since crystallization is slow, allow plenty of time for it.

7. This and similar references are to the techniques described in Sec. 3.3.

8. Note that the lake is a blue *precipitate*. The solution itself should be purple. If there is any doubt as to the color of the precipitate, remove the solution (T-9) and wash the precipitate once with 5–10 drops of water (T-10). Recentrifuge and observe the precipitate. Avoid an excess of S. and O. reagent, for it has a very strong color and may make the lake too dark.

9. If the solution starts to spatter, remove it from the flame for a few seconds. Keep the casserole well above the flame and swirl the contents steadily. Patience is required.

10. Failure to remove ammonium salts completely will lead to a false test for potassium, but if too high a temperature is used, the alkali metal salts may fuse with the glaze of the dish and be lost.

11. These tests are based on formation of fairly soluble precipitates and require concentrated solutions and excess of reagents. If you add too much water here, you may fail to get the tests.

12. The sodium cobaltinitrite reagent is unstable. It should have a deep reddish amber color. Pale yellow or pink solutions should be discarded. Test the reagent with a standard solution of K^+ if its potency is doubted.

13. A voluminous precipitate at this point is often $(NH_4)_2NaCo(NO_2)_6 \cdot H_2O$, which is indistinguishable from the potassium compound, if ammonium salts were not completely removed. Warming will remove a small amount of the ammonium compound but not a large quantity.

14. Dissolve the precipitate in one drop of 6 M HCl. View the flame through two thicknesses of cobalt blue glass to cut out the sodium color. The potassium flame appears red.

15. Both the potassium and the sodium precipitates supersaturate easily. See Note 6.

EXERCISES

11.1. Name the ions of this analytical group. Why are they grouped together?
What are the positions of the metals in the periodic table?

11.2. Write equations for the reactions of these ions that are used in the analytical
procedure.

11.3. Explain why magnesium hydroxide is not precipitated with Cation Group 3
and magnesium carbonate with Cation Group 4.

11.4. What reagent or combination of reagents or what operation would you
use to separate in one step the two substances in each pair?

(a) Mg^{++}, Ca^{++} (b) NaCl, AgCl

(c) K^+, Mg^{++} (d) NH_4Cl, KCl.

11.5. Give two reasons why an ammonium salt must be present when magnesium
ammonium phosphate is precipitated.

11.6. Why must ammonium salts be removed before the potassium test? How
does each of the following accomplish this?
(a) Heating the solid residue.
(b) Evaporating with nitric acid.
(c) Heating with sodium nitrite.
(d) Evaporating with sodium hydroxide.

11.7. What mistakes could lead to the following errors?
(a) A large yellow precipitate is obtained in the K^+ test, although potassium is not present.
(b) Very pale yellow needles form upon addition of magnesium uranyl acetate reagent, and yet the flame test is violet.
(c) Sodium is present, and yet no precipitate is obtained.
(d) Potassium is present, but a pink solution is obtained and no precipitate.
(e) Ammonium ion is absent, and yet pink litmus over the alkaline solution
turns blue immediately
(f) Ammonium ion is absent, and yet a test for it is obtained on Solution
6.1.

11.8. How would you distinguish between the two solids in each pair?

(a) $(NH_4)_2SO_4$, K_2SO_4 (b) $MgCl_2$, KCl

(c) KNO_3, $NaNO_3$ (d) $Mg(OH)_2$, NaOH

11.9. Calculate the number of milligrams of Na^+ required to give 50 mg of
$NaZn(UO_2)_3(OAc)_9 \cdot 9H_2O$.

11.10. Calculate the equilibrium constant for $NH_4Cl(s) \leftrightharpoons NH_3(g) + HCl(g)$ at
25°C from the following standard Gibbs energies of formation:

$$\Delta G_f°(NH_3) = -3.976; \quad \Delta G_f°(HCl) = -22.77; \quad \Delta G_f°(NH_4Cl)$$
$$= -48.73 \text{ kcal mol}^{-1}$$

What is the significance of the result?

TWELVE

Analysis of Alloys

and General Cation Unknowns

12.1 Preparation of a Solution of an Alloy. Principles

Alloys are intimate mixtures of metals; they may be compounds, solid solutions, or heterogeneous mixtures. The properties of a metal can be considerably altered by alloying—even its chemical reactivity may change. Alloys also sometimes contain nonmetals such as carbon, silicon, phosphorus, and arsenic. The first two are often added to impart desirable properties such as hardness or resistance to corrosion; the others are usually undesirable and are present as impurities.

Acids are the most commonly used solvents for metals and alloys. Hydrochloric acid is a satisfactory solvent for only the more active metals such as aluminum or magnesium, for the hydrogen ion is not a powerful oxidizing agent:

$$2H^+ + 2e^- \rightarrow H_2 \qquad E^\circ = 0.00 \text{ V}$$

It is not even used to dissolve active metal alloys, for these may contain small amounts of an inactive metal that would then escape detection. Furthermore, if arsenic or antimony were present, it would be reduced to the hydride and be partially lost:

$$Al + As + 3H^+ \rightarrow AsH_3(g) + Al^{3+}$$

Dilute (6–8 M) nitric acid is the most generally satisfactory solvent for alloys.

The nitrate ion in acid solution is a powerful oxidizing agent

$$4H^+ + NO_3^- + 3e^- \rightarrow NO(g) + 2H_2O \qquad E^\circ = 0.96 \text{ V}$$

and will therefore oxidize inactive metals such as copper ($E^\circ = 0.34$ V) or silver ($E^\circ = 0.80$ V). The dilute acid is usually preferred over the concentrated (16 M) acid for two reasons. (1) Several nitrates—Pb(NO$_3$)$_2$ for one—are only slightly soluble in the concentrated acid and form surface coatings on the unreacted metal that slow down reaction. (2) Several metals—notably Cr, Al, and Fe—are passive towards the concentrated acid and virtually do not dissolve in it. This is attributed to the formation of an impervious oxide film on the surface of the metal that prevents further attack by the acid. The principal problem with the use of nitric acid as a solvent is its action on tin and antimony. These elements are converted to hydrated oxides, $SnO_2 \cdot xH_2O$ (metastannic acid) and $Sb_2O_5 \cdot yH_2O$, which are difficult to bring into solution.

Next to dilute nitric acid, aqua regia is the most satisfactory solvent for alloys. It combines the oxidizing action of nitric acid with the formation of chloro complexes and is also a more rapid oxidizing agent than nitric acid alone. The oxides of tin and antimony are not obtained, but silver and lead are converted to chloro complexes and insoluble chlorides. Typical reactions of metals with aqua regia are represented by

$$2H^+ + NO_3^- + Cl^- + Ag(s) \rightarrow AgCl(s) + NO_2(g) + H_2O$$

$$4H^+ + 2NO_3^- + 4Cl^- + Hg(l) \rightarrow HgCl_4^{--} + 2NO_2(g) + 2H_2O$$

$$5H^+ + 5NO_3^- \qquad + As(s) \rightarrow H_3AsO_4 + 5NO_2(g) + H_2O$$

$$8H^+ + 4NO_3^- + 6Cl^- + Sn(s) \rightarrow SnCl_6^{--} + 4NO_2(g) + 4H_2O$$

$$4H^+ + 2NO_3^- + 4Cl^- + Cu(s) \rightarrow CuCl_4^{--} + 2NO_2(g) + 2H_2O$$

12.2 Preparation of a Solution of an Alloy. Procedure

The alloy must be in finely divided form: powder, shavings, turnings, drillings, or parings.

(a) Preliminary Solubility Tests

Test the solubility of very small portions (5–10 mg) in 8 M HNO$_3$ (dilute the concentrated acid with an equal volume of water). Heat in the water bath until reaction appears to be complete. Centrifuge and examine any residue. A dark residue is unreacted metal or carbon; remove the solution and try to dissolve the residue in a fresh portion of nitric acid. If it does not dissolve, test the solubility of another very small portion of the alloy in aqua regia. White residues may form with either solvent; disregard these for the time being and look for dark particles of unreacted metal. If neither solvent appears to dissolve your alloy, consult your instructor. Other solvents such as a combination of bromine and hydrochloric acid are useful for particular types of alloys.

(b) Dissolution of the Alloy

Weigh 50 mg of alloy into a test tube or casserole and add a small amount of the solvent that has been selected. Heat the test tube in vigorously boiling water or the casserole over a microflame (do not boil). Stir frequently and break up clumps of particles or rub off surface coatings to expose fresh surfaces. Add more acid if vigorous reaction has ceased and dark particles of metal remain, but avoid a large excess of acid, for this must be removed by evaporation in a subsequent step. Bear in mind that *many alloys dissolve slowly at best*, and allow plenty of time for reaction to occur. Centrifuge and withdraw the clear solution to a casserole; its treatment is described in step (c). Examine the residue.

(1) If the residue is white, it may be oxides of tin and antimony or the chlorides of silver and lead, depending on whether nitric acid or aqua regia was used as the solvent. Treat it according to step (c).

(2) If the residue is dark, it may contain unreacted metal or it may be carbon. Try to dissolve it in a fresh portion of solvent. Should this fail, it may be carbon or metal in a passive state. Wash the residue two or three times with water to remove nitrate, and try to dissolve it in 6 M or 12 M HCl. Ordinarily, any *small* amount of black residue that resists nitric acid, aqua regia, and hydrochloric acid can be regarded as carbon and discarded.

(c) Preparation of the Solution for Cation Analysis

Transfer the solution with any white residue to a casserole and evaporate it over a microflame to a volume of a few drops. Withdraw the flame and let the heat of the dish complete the evaporation. If there is a white residue of the oxides of tin and antimony, it is advisable to coagulate them further by a second evaporation with 6 M HNO_3. Dissolve the salts in 10 drops of 6 M HNO_3 and warm briefly to hasten their solution. Transfer the mixture to a centrifuge tube and centrifuge. Withdraw the solution to a test tube. Wash the residue twice with water containing a little nitric acid and combine the washings with the solution. Mix thoroughly and use one-third to one-half of the solution for the systematic analysis (Sec. 12.3).

(d) Treatment of the White Residue from Step (c)

(1) If nitric acid was used as the solvent, the residue may contain hydrated forms of SnO_2, Sb_2O_5, and SiO_2 together with other occluded metal ions. Add 5 drops of 6 M NH_3 and 2 drops of TA solution and warm in the hot water bath. Centrifuge and withdraw the solution to another test tube. Repeat the extraction of the residue. Acidify the combined extracts with 2 M HCl to reprecipitate the sulfides of antimony and tin. Dissolve these sulfides in 12 M HCl as indicated under Precipitate 4.1, Outline 4, Chapter 8, and test for antimony and tin according to that procedure. If a black residue remains after the extraction of antimony and tin as thio anions, it will be sulfides of other cations. Dissolve it in nitric acid and add the solution to that obtained in step (c).

(2) If aqua regia was used as the solvent, a white residue is probably AgCl or $PbCl_2$. Analyze it by Outline 1, Chapter 7.

12.3 Analysis of General Cation Unknowns

The sample may be a solution of an alloy [Sec. 12.2(c)] or a "general unknown" to be analyzed only for cations, the anions being nitrate or chloride.

If an alloy is being analyzed, it will be unnecessary to test for NH_4^+, Fe^{++}, or Hg_2^{++}. Assume further, unless directed otherwise, that Na^+, K^+, Ca^{++}, Sr^{++}, and Ba^{++} are absent. Thus Mg^{++} ion is the only member of Groups 4 and 5 to be sought.

(a) Note the color of the solution and draw what conclusions you can from it [Table 6.1 and Sec. 6.3(a)]. Bear in mind that if the solution is colorless, certain ions are absent.

(b) Make a few crash precipitation tests [Secs. 6.4(a)–(c)].

(c) Tests for the following cations are made on the original solution: Fe^{++}, Fe^{3+}, Ni^{++}, Mn^{++}, Co^{++}, Zn^{++} (Sec. 9.15, Cation Group 3), and NH_4^+ (Cation Group 5). Flame tests may also be done on this solution. All such tests should be confirmed by a systematic analysis.

(d) To make a systematic analysis for the cations, follow the procedures in Outlines 1–7. Three features of the general analysis for cations of all groups must be emphasized:

(1) Always test for completeness of precipitation. Failure to do this may lead not only to the loss of ions but also to confusion in later tests when foreign ions leak through and interfere.

(2) Wash precipitates thoroughly. Use a rod to disperse the precipitate in the wash water. Large gelatinous precipitates of hydroxides or sulfides are often contaminated with considerable amounts of unwanted ions.

(3) Adjust the volumes of reagents for precipitates or amounts of ions of abnormal size. Frequently, in the analysis of alloys, a few metals will be present in large amounts, and the quantities of reagents must be increased accordingly.

(e) An alternative separation of Cation Group 4 is advantageous, for these ions may be lost in a systematic analysis by precipitation with earlier groups. The slow oxidation of hydrogen sulfide or thioacetamide to sulfate by oxygen of air or nitric acid will result in the loss of barium or strontium. Because of this, thioacetamide and hydrogen sulfide are destroyed by evaporation with concentrated hydrochloric acid after precipitation of Cation Groups 2 and 3. Old solutions of thioacetamide should not be used as reagents if they give any turbidity with barium chloride. The carbonates of Group 4 may prematurely precipitate with Group 3 if the ammonia reagent has absorbed carbon dioxide from the air. Other anions such as phosphate or fluoride also cause precipitation of Group 4 with Group 3. Even when such anions are absent, the ions of Group 4 may be co-precipitated with large gelatinous precipitates of aluminum or chromium(III) hydroxides. Should the precipitate of the carbonates of Group 4 be so small as a result of these losses that satisfactory tests cannot be obtained, it is desirable to use an alternative separation.

The precipitation of the sulfates of Group 4 is an attractive separation, for it can be made in acidic solution before the precipitation of Group 2, in other words, before the losses occur. Lead sulfate, to be sure, will also precipitate, but it is easily separated. Calcium sulfate is only sparingly soluble, but its solubility can be decreased by addition of ethyl alcohol, which has a lower dielectric constant than water. Lead sulfate can be dissolved by forming the acetate or hydroxide complex of lead. The latter works better because sodium carbonate can be added with the excess of sodium hydroxide to transpose the sulfates of calcium and strontium into the less soluble carbonates. Typical reactions are represented by

$$PbSO_4(s) + 3OH^- \rightarrow Pb(OH)_3^- + SO_4^{--}$$

$$CaSO_4(s) + CO_3^{--} \rightarrow CaCO_3(s) + SO_4^{--}$$

Transposition of barium sulfate to the carbonate is more difficult but usually occurs to a sufficient extent to permit the detection of barium. The procedure for the sulfate separation is given in Outline 8.

OUTLINE 8. The Sulfate Separation of Cation Group 4

Solution containing all cations but Ag^+ and Hg_2^{++}. Total vol should be 0.5 ml (10 drops); adjust by evapn or diln. Add 1 drop 18 M H_2SO_4 and 20 drops 95% C_2H_5OH. Stir vigorously for 1 minute.[1] Centrifuge and wash ppt twice with soln made by mixing 10 drops water, 1 drop 18 M H_2SO_4, and 10 drops C_2H_5OH.

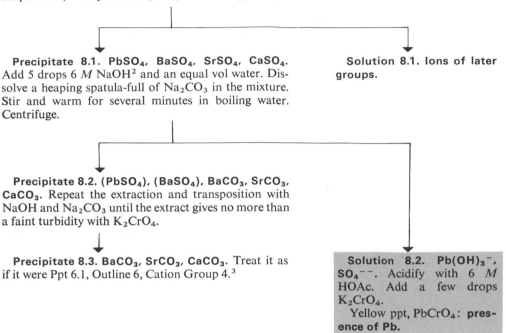

Precipitate 8.1. $PbSO_4$, $BaSO_4$, $SrSO_4$, $CaSO_4$. Add 5 drops 6 M $NaOH$[2] and an equal vol water. Dissolve a heaping spatula-full of Na_2CO_3 in the mixture. Stir and warm for several minutes in boiling water. Centrifuge.

Solution 8.1. Ions of later groups.

Precipitate 8.2. ($PbSO_4$), ($BaSO_4$), $BaCO_3$, $SrCO_3$, $CaCO_3$. Repeat the extraction and transposition with NaOH and Na_2CO_3 until the extract gives no more than a faint turbidity with K_2CrO_4.

Precipitate 8.3. $BaCO_3$, $SrCO_3$, $CaCO_3$. Treat it as if it were Ppt 6.1, Outline 6, Cation Group 4.[3]

Solution 8.2. $Pb(OH)_3^-$, SO_4^{--}. Acidify with 6 M HOAc. Add a few drops K_2CrO_4.
Yellow ppt, $PbCrO_4$: **presence of Pb.**

Notes on Outline 8

1. Calcium sulfate supersaturates; hence the vigorous stirring over a period of time. It is advisable to let Solution 8.1 stand for 5–10 minutes in case more calcium sulfate precipitates.

2. If lead is known to be absent from the regular systematic analysis, the sodium hydroxide may be omitted.

3. Transposition of barium sulfate may be incomplete, so that all of Precipitate 8.3 may not dissolve in acetic acid. Usually a sufficient amount of barium is brought into solution to give a precipitate with chromate. If there is a large residue insoluble in acetic acid, it can be dried and fused with sodium carbonate. (See E. J. King, *Qualitative Analysis and Electrolytic Solutions*, Harcourt Brace Jovanovich, New York, 1959, Sec. 24–8c(3), pp. 571–572.)

EXERCISES

12.1. Outline the simplest scheme for analysis of each of the following alloys. What solvent would be used for each?
(a) Wood's metal: Bi, Pb, Sn, and Cd.
(b) Dow-metal C: Al, Mg, and Zn.
(c) Nickel silver: Cu, Zn, and Ni.
(d) Chromel: Ni, Fe, and Cr.

12.2. Why is it unnecessary to test for NH_4^+, Fe^{++}, or Hg_2^{++} ions in solutions of alloys?

12.3. Write equations for the action of nitric and hydrochloric acids separately on these metals: Al, Fe, Sn, Cu, Zn, As, Mg.

12.4. Summarize the advantages and disadvantages of hydrochloric acid, nitric acid, and aqua regia for dissolving alloys.

12.5. Summarize the ways in which ions of Group 4 can be lost during a systematic analysis, and give the theoretical basis for the sulfate separation of Outline 8.

12.6. Explain the theoretical basis for separations in the cation analysis that depend on control of the concentration of (a) S^{--}, (b) OH^-, (c) CO_3^{--}, (d) CrO_4^{--}.

12.7. Cite examples from the systematic analysis for cations, exclusive of group separations, of (a) chelation, (b) common ion effect, (c) formation of ammonia complexes, (d) amphoterism, (e) oxidation or reduction.

THIRTEEN

Analysis of Mixtures

for Anions

13.1   Introduction

The abbreviated version of anion analysis that was given in Chapter 4 can be used only when a single anion is present. With mixtures containing two or more anions, interferences must be considered in making tests. Similarity in behavior is one cause of interference. We cannot get a definite test for chloride by precipitating silver chloride if bromide and iodide are present, for they too form insoluble silver salts. A systematic procedure is therefore required to eliminate Br^- and I^- ions before testing for Cl^- ion; it is given in Sec. 13.3. A second cause of interference is direct reaction between two ions. One may be a reducing agent like S^{--} or I^-, the other an oxidizing agent like SO_3^{--} or NO_2^-. Usually no reaction occurs in neutral or basic solution, but when the solution is acidified the two ions destroy each other:

$$H_2SO_3 + 2H_2S \rightarrow 3S(s) + 3H_2O$$

$$2H^+ + 2HNO_2 + 2I^- \rightarrow I_2 + 2NO(g) + 2H_2O$$

The elimination tests given in Chapter 4, with the exception of ET-5, can be used for mixtures as well as for simple salts. The identification tests for the individual ions will usually need to be modified to take into account the possible presence of interfering ions.

13.2 Preparation of a Solution for Analysis

The directions in Sec. 4.2 are applicable. If a residue insoluble in acid remains after the sodium carbonate treatment, it may contain barium sulfate, silver halides, or insoluble sulfides that are difficult to transpose with carbonate or naturally occurring fluorides, phosphates, and carbonates. Provision is made for such residues in the identification tests.

13.3 Elimination Tests

Tests ET-1, 2, 3, 4, 6, and 7 can be performed as described in Chapter 4. The principles on which they are based are treated in Chapter 5.

Modified Elimination Test ET-5 for Anions that Form Insoluble Silver Salts

Although a large number of anions form silver salts that are insoluble in water or neutral solutions, most of them are dissolved by a strong acid or ammonia. By regulating the concentration of ammonia (Sec. 2.10), which in turn controls the silver ion concentration through the equilibrium

$$Ag^+ + 2NH_3 \leftrightharpoons Ag(NH_3)_2{}^+$$

it is possible to precipitate the least soluble silver salts—the sulfide, chloride, bromide, and iodide—but not the rest.

Procedure: To 2 drops of 3.3 M AgNO$_3$ (solution A at side shelf), add ammoniacal ammonium carbonate (solution B at side shelf) drop by drop until the precipitate that forms at first just dissolves. Avoid an excess; about 5–7 drops of solution B should be required.

To the reagent you have just prepared, add 20 drops of the solution under test. Mix well and allow to stand for a minute. If no precipitate forms, the absence of Cl$^-$, Br$^-$, and I$^-$ is indicated. Centrifuge. Draw off the solution and discard it. Wash the precipitate once with water. If the wash water becomes hazy due to the presence of colloidal silver salts, add a drop of NH$_4$NO$_3$ solution to coagulate the colloid.

A black precipitate at this stage indicates the *presence of sulfide*, but it may be difficult to discern in a large precipitate of the other silver salts. To confirm, transfer a little of the precipitate to a small square of filter or spot paper with a pipet. Add a few drops of 0.5 M Na$_2$S$_2$O$_3$ solution to dissolve the silver halides; for example,

$$AgI(s) + 2S_2O_3{}^{--} \rightarrow Ag(S_2O_3)_2{}^{3-} + I^-$$

A residual brown or black stain on the white paper confirms the presence of sulfide. If insoluble sulfides were not transposed with carbonate in making the Prepared Solution, no test for sulfide will be obtained here.

Silver must now be removed from the residual silver salts. Add to the precipitate 20 drops of 6 M HOAc and half a spatula-full of zinc metal. Heat the tube in the water bath for 15 minutes. Break apart the clumps of precipitate with a rod occasionally to bring fresh surfaces in contact with the reagent. Add more zinc if necessary to ensure a continuous reaction. Zinc reduces silver ion to the metal, releasing the anions:

$$Ag_2S(s) + Zn(s) + 2H^+ \rightarrow 2Ag(s) + Zn^{++} + H_2S(g)$$

$$2AgBr(s) + Zn(s) \rightarrow 2Ag(s) + Zn^{++} + 2Br^-$$

While the reaction proceeds, carry out the remaining elimination tests and prepare the elimination chart (Sec. 13.5). At the end of 15 minutes, dilute the mixture with 5–10 drops of HOAc and transfer it to a centrifuge tube. Centrifuge and transfer the solution to a test tube; save it for the analysis for Cl^-, Br^-, and I^- [Sec. 13.4(b)].

If the original sample contains a silver halide, it may not have been transposed by sodium carbonate in making the Prepared Solution (Sec. 4.2). The halide can be brought into solution with zinc and acetic acid by the procedure used here.

13.4   Identification Tests

(a) Identification of Fluoride Ion

The procedure given in Sec. 4.5 is still applicable.

(b) Separation and Identification of Chloride, Bromide, and Iodide Ions

Start with the acetic acid solution containing the ions that was obtained in the modified ET-5 (Sec. 13.3). Selective oxidation is used to separate and identify the ions (Sec. 2.12).

(1) *Identification of Iodide.* Add several crystals of sodium nitrite and several drops of CCl_4 to the solution under test. Shake it vigorously to extract iodine. The *presence of iodide* is indicated by a violet lower layer:

$$2HNO_2 + 2H^+ + 2I^- \rightarrow 2H_2O + 2NO(g) + I_2$$

(2) *Removal of Iodide.* Transfer the water layer from the identification step to a casserole and add about one-half spatula-full of sodium nitrite. Evaporate the solution to dryness (T-7) under a ventilating duct or in a hood. Add 10 drops of 6 M HOAc and more sodium nitrite and evaporate to dryness again.

After the last evaporation, dissolve the residue in 10 drops of water and add 2 drops of 6 M HNO_3. Transfer the solution to a test tube and add a few crystals of sulfamic acid until evolution of nitrogen ceases. Add a few more crystals and see if more gas is formed. Continue until no more nitrogen is evolved. The excess of nitrite is then destroyed:

$$HNO_2 + NH_2SO_3^- \rightarrow N_2(g) + HSO_4^- + H_2O$$

(3) *Identification of Bromide.* Add 5 drops of 6 M HNO_3 to the solution, and then add 2 drops of CCl_4. (Avoid using too much CCl_4 because the color of bromine is more intense in a smaller volume.) Now add 0.02 M $KMnO_4$ drop by

drop, shaking after each addition, until the water layer remains pink for at least a minute or an orange bead is obtained. The orange lower layer indicates the *presence of bromide:*

$$2MnO_4^- + 16H^+ + 10Br^- \rightarrow 5Br_2 + 2Mn^{++} + 8H_2O$$

(4) *Removal of Bromide.* If bromide was present, transfer the solution to a casserole. Add 5 more drops of 6 M HNO_3 and sufficient 0.02 M $KMnO_4$ to give a purple color that persists for about a minute. Evaporate the solution to half its initial volume to drive off bromine. Then dissolve any brown MnO_2 and reduce excess $KMnO_4$ with a drop of 3% H_2O_2:

$$2MnO_4^- + 3Mn^{++} + 2H_2O \rightarrow 5MnO_2(s) + 4H^+$$

$$MnO_2(s) + H_2O_2 + 2H^+ \rightarrow O_2(g) + Mn^{++} + 2H_2O$$

(5) *Identification of Chloride.* Add a drop of $AgNO_3$ solution to the bromide-free solution. Centrifuge and discard the solution. Wash the precipitate, if any, once or twice with water to remove Mn^{++} ions. Add a few drops of 6 M NH_3 to the precipitate. If it is silver chloride, it should dissolve. Acidify the solution with 6 M HNO_3 (T-12). The *presence of chloride* is indicated by white, curdy AgCl. A hazy opalescence that does not centrifuge down is caused by a trace of chloride and should not be reported.

(c) Identification of Sulfide Ion

The elimination tests ET-1, 2, 3, and 5 give reliable indication of the *presence of sulfide.* The modified ET-5 (Sec. 13.3) separates sulfide from oxidizing agents that might destroy it. When the sodium carbonate treatment (Sec. 4.2) leaves a residue insoluble in acetic acid, treat a portion of it with a few granules of zinc and a few drops of 6 M H_2SO_4. Detect hydrogen sulfide gas with moist lead acetate paper:

$$CuS(s) + Zn(s) + 2H^+ \rightarrow Zn^{++} + Cu(s) + H_2S(g)$$

$$Pb(OAc)_2 + H_2S \rightarrow PbS(s) + 2HOAc$$

This can be done on the original sample too if sulfite is absent; sulfite is also reduced to H_2S by zinc and acid.

(d) Identification of Sulfite and Sulfate Ions

Sulfite is identified by its reducing action on permanganate [Sec. 4.5(f)], but other reducing agents such as S^{--} or NO_2^- will also decolorize the reagent. Unlike sulfite, they do not form barium salts that are insoluble in neutral or basic solutions. Oxidizing agents such as NO_2^- and NO_3^- also interfere, because they convert sulfite to sulfate in acid solution. No decolorization of permanganate will then occur, but barium sulfate will still form if barium ion is present.

Treat several drops of the Prepared Solution with sufficient $BaCl_2$ solution to precipitate the insoluble barium salts. Wash the precipitate three times with hot water; add about 10 drops of water each time, heat in the water bath for a minute, and centrifuge. Discard the washings, which contain interfering ions. Add a few drops of 6 M HCl and dilute to a convenient volume with about 10 drops of water. The mixture should be acid. A white residue of $BaSO_4$ insoluble in HCl indicates the *presence of sulfate*. Remember, though, that sulfate is frequently present as a result of oxidation of sulfite by oxygen of air. The residue may range from a faint turbidity to an appreciable precipitate, depending on circumstances. If sulfite is present and the $BaSO_4$ precipitate is small, compare it with one obtained by using one drop of a solution prepared by diluting 1 to 10 the standard solution of sulfate ion. Report sulfate only if the bulk of your precipitate is larger than this control.

Centrifuge to settle the barium sulfate and withdraw the solution to a clean tube. Add one drop of 0.02 M $KMnO_4$. Decolorization of the permanganate and formation of white $BaSO_4$ are indications of the *presence of sulfite*. Slow decolorization of permanganate will occur if the chloride concentration is high, but no barium sulfate will form then.

(e) Identification of Nitrite Ion

Carbonate and sulfate interfere with the sulfamic acid–nitrous acid reaction [Sec. 4.5(h)], the first by forming bubbles of CO_2 that could be confused with N_2, the second by precipitating barium sulfate. Both can be removed by precipitation with barium chloride.

To a few drops of the Prepared Solution, add $BaCl_2$ solution until precipitation is complete. Centrifuge and transfer the solution to a test tube. Add a few crystals of solid sulfamic acid and a drop of $BaCl_2$ solution. Flick the bottom of the tube with your forefinger to cause vigorous evolution of gas bubbles. The formation of both a white precipitate and a gas indicates the *presence of nitrite*.

(f) Identification of Nitrate Ion

The test in Sec. 4.5(i) is based on the reduction of nitrate to ammonia. Nitrite undergoes the same reaction and must be removed. Even then, nitrites are frequently contaminated by nitrate formed by slow oxidation by oxygen of air, so that if nitrite is present, a small test for nitrate is usually obtained. Ammonium salts must also be absent, for they liberate ammonia in basic solution.

If nitrite has been found to be present, add several crystals of sulfamic acid to 5 drops of the Prepared Solution in a test tube and warm until bubbles of nitrogen are no longer evolved. The solution should be acidic. Add a little more sulfamic acid to check the completeness of removal of nitrite. Now carry out the procedure described in Sec. 4.5(i), including removal of ammonium ion prior to the reduction.

(g) Identification of Phosphate Ion

High concentrations of chloride or sulfate retard the precipitation of ammonium molybdophosphate [Sec. 4.5(j)], and reducing agents destroy molybdate. Some phosphates are not readily transposed by sodium carbonate but are dissolved by nitric acid. This solvent also destroys chloride and reducing agents.

Treat 20–40 mg of the original sample with several drops of 16 M HNO$_3$. Warm in the water bath for a few minutes, then dilute with several drops of water, centrifuge, and remove the clear solution to a clean tube. Add 2 drops of ammonium molybdate reagent to the acidic solution and allow it to stand for 5 minutes. The *presence of phosphate* is indicated by the formation of a bright yellow precipitate of $(NH_4)_3[PMo_{12}O_{40}] \cdot 6H_2O$. A white precipitate of MoO_3 may form if phosphate is absent or if the tube is too hot (above 50°C, too hot to hold).

(h) Identification of Carbonate Ion

Sulfites as well as carbonates react with barium hydroxide solution to give a white turbidity. They are removed by oxidation to sulfate with hydrogen peroxide before the addition of acid. Nitrite is likewise oxidized:

$$HO_2^- + SO_3^{--} \rightarrow SO_4^{--} + OH^-$$

$$HO_2^- + NO_2^- \rightarrow NO_3^- + OH^-$$

In the presence of nitrites and sulfites, treat 20–30 mg of the solid sample (*not* the Prepared Solution) with several drops of 3% H$_2$O$_2$. Warm for several minutes in the hot water bath; then carry out the test described in Sec. 4.5(k).

Table 13.1

A Sample Elimination Chart

Ion	ET-1	ET-2	ET-3	ET-4	ET-5	ET-6	ET-7	Summary	Confirmation
F$^-$						A		A	
Cl$^-$				?				?	P
Br$^-$				?				?	A
I$^-$			A		(A)			A	
S^{--}	?	A	A		A			A	
SO$_3^{--}$	?	?	A				(A)	A	
SO$_4^{--}$							P	P	
NO$_2^-$		A	A	(A)				A	
NO$_3^-$				P				P	P
PO$_4^{3-}$	?					A		A	
CO$_3^{--}$	?	?						?	P

13.5   Analysis of a Known Mixture

For practice, try the elimination and identification tests on a mixture that contains 1 part KBr, 1 part KI, 2 parts anhydrous Na_2SO_3, 2 parts KNO_3, 2 parts Na_3PO_4, and 2 parts Na_2CO_3 by weight. Prepare a solution of the unknown by dissolving 0.10 g of it in 5 ml of distilled water. Even though Cl^- ion is absent, test for it to see if removal of bromide and iodide was complete.

Tabulate the results of the elimination tests in a chart of the form indicated in Table 13.1. The form in the table is filled out for a mixture that is different from the one assigned for practice but one that is also water soluble and containing no heavy metals. In the chart, the letter A stands for absent, P for present, (A) for absent when this was indicated by earlier tests, and ? when the test gives insufficient information to permit a distinction between several possible ions. The most significant entry is A; no further test is required. Although an ion appears to be present, it may be advisable to confirm this by a further test.

EXERCISES

13.1.   Why is an alkaline, rather than an acidic, solution of the unknown used for the tests?

13.2.   Explain why a combination of silver nitrate, ammonia, and ammonium carbonate is used in ET-5.

13.3.   A certain unknown was completely soluble in water and contained no heavy metals. In ET-1 the pH was found to be 11. In ET-2 a red brown gas was evolved. In ET-3 a deep blue precipitate was formed. In ET-4 there was a yellow color. In ET-5 no precipitate was obtained. In ET-6 a white precipitate formed upon addition of $Ca(NO_3)_2$ solution. An acidic solution withdrawn from above the precipitate gave no visible reaction when it was neutralized with ammonia. In ET-7 no precipitate was obtained. On the basis of these observations, draw up an elimination chart for this unknown.

13.4.   Why are the following pairs incompatible in acid solution? Write equations for the reactions: (a) SO_3^{--}, NO_2^-;   (b) NO_2^-, S^{--}.

13.5.   Why are the barium salts precipitated in alkaline solution before testing for sulfite and sulfate?

13.6.   What error would account for the following observations?
(a) A yellow precipitate in the chloride test.
(b) A white precipitate in the phosphate test.
(c) A turbidity in the carbonate test that dissolved in nitric acid and decolorized 0.002 M $KMnO_4$.

13.7.   No test was obtained for sulfide from an acid solution of the sample, yet a black precipitate of lead sulfide was obtained by adding lead acetate to an alkaline solution. Suggest an explanation.

13.8.   Write equations for the reactions used in the final identification of NO_2^-, SO_3^{--}, CO_3^{--}, I^-, and S^{--} ions.

13.9.   In each of the following pairs the second ion interferes with the test for the first. Explain why it does so and how the interference can be removed.

(a) CO_3^{--}, SO_3^{--} (b) Cl^-, I^- (c) NO_3^-, NO_2^-

FOURTEEN

Analysis of Mixtures

for Cations and Anions

14.1 Preliminary Examination

Note the obvious physical characteristics of the material. If it is colored, consider whether any deductions can be made with the aid of Table 6.1. If it is colorless, certain constituents are eliminated. Examine it closely, preferably with a magnifying glass, for evidence of heterogeneity. It may be possible to estimate from this observation the minimum number of components of the mixture.

(a) Preparation of the Sample

The sample taken for analysis must be representative of the entire lot of material. Since we use 100 mg or less for analysis, sampling can be something of a problem when a large quantity of material is available. The most formidable difficulties involved in sampling will probably not confront you in this course; for insight into the problems, consult one of the larger textbooks of quantitative analysis.

The material to be analyzed must be finely powdered and intimately mixed. If it is not—for example, if the particles of the individual components can be distinguished—grind it to an impalpable powder.

(b) Flame Tests

Review the discussion in Secs. 6.2(b), 10.8, and 11.10; then moisten your sample with hydrochloric acid and try the flame test on it.

(c) Solubility Tests

This subject has previously been discussed in Secs. 3.3 (T-2) and 6.3(a). Test the solubility of small samples (about 10 mg) of the material in water, 6 M HCl, and 6 M HNO$_3$. Warm in each case if nothing happens in the cold solvent. Watch for evolution of a gas with water or HCl. Consult the extended Summary of Solubilities in Appendix A.1 or a chemical handbook and see if certain classes of substances or even specific substances can be eliminated or tentatively identified. If the unknown is completely soluble in water, for example, it may contain sodium, potassium, or ammonium salts of various anions or nitrates, acetates, or nitrites of various cations. If it is insoluble in water and hydrochloric acid but soluble in nitric acid, it does not contain insoluble sulfates but may contain sulfides or silver salts of weak acids. It must be noted that certain naturally occurring substances are less soluble than the same material freshly precipitated, and certain salts such as lead chromate and various sulfates and oxides are much less soluble after they have been heated strongly.

(d) Other Preliminary Tests

Consult Secs. 6.4(a)–(c) for various crash tests that are worth trying because they may give some leads as to the presence or absence of groups of ions.

14.2   Analysis for Anions

It is usually convenient to analyze for anions before cations, although no rigid rule need be established. Some anions interfere in the separations of cation analysis, and their presence should be known before the systematic analysis of the cations is attempted. Certain anions and cations (for example, Ag$^+$ and Cl$^-$) are also incompatible in solution [Sec. 6.3(c)]; and if the presence of one ion is established, the other is necessarily absent from a solution of the sample.

The Prepared Solution for anion analysis should be prepared as described in Sec. 4.2. Then, elimination tests ET-1, 2, 3, 4, 6, and 7 in Chapter 4 and ET-5 in Sec. 13.3 should be carried out and the elimination chart (Table 13.1) constructed. Finally, such identification tests as may be required should be carried out according to the directions in Chapter 13.

14.3   Preparation of a Solution for Cation Analysis

If the solubility tests of Sec. 14.1 were inconclusive, test the solubility of separate small portions of the sample successively in cold and hot portions of

$$H_2O, 6 \ M \ HCl, 12 \ M \ HCl, 6 \ M \ HNO_3, 16 \ M \ HNO_3, \text{ and aqua regia}$$

Use 5–10 drops of each solvent, stir it thoroughly with about 10 mg of sample, and allow time for reaction to occur. Centrifuge to determine if a small residue remains undissolved. When concentrated acids are used, dilute the mixture slowly with an equal volume of water after reaction ceases. Certain salts are only sparingly soluble in the concentrated acids.

Weigh 100 mg of the sample and treat it with the solvent you selected. If no solvent dissolved the sample completely, use the one that had the most effect or, if there is little choice, use aqua regia. Directions are given in Sec. 14.4 for treatment of insoluble residues.

When the solution was obtained by use of one of the acid solvents, evaporate it to a volume of a few drops to remove excess acid. Do not evaporate to dryness, for this may cause decomposition of formation of basic salts that are difficult to redissolve. Take about one-fourth of the solution (the equivalent of 25 mg of the original sample) for the systematic analysis for the cations. Refer to Sec. 12.3 for suggestions about general cation analysis.

14.4   Treatment of Insoluble Residues

If there is a residue insoluble in acid, special treatment is required. The common substances the residue may contain are

$PbSO_4$, $BaSO_4$, $SrSO_4$, $CaSO_4$ (if a large amount is present), ignited sulfates of Fe(III), Al(III), or Cr(III), $PbCrO_4$ (fused)

AgCl, AgBr, AgI, CaF_2 (the mineral form)

Al_2O_3, Cr_2O_3, SnO_2, Sb_2O_5, SiO_2, or silicates

C or S

The residue is treated first with sodium carbonate solution to transpose the sulfates, then with zinc and acid to reduce the silver halides.

(a) Wash the residue twice with a moderately dilute solution of the acid used to dissolve the sample. Then suspend it in 1 ml of water and add two heaping spatulas-full of sodium carbonate. Heat in boiling water for 10 minutes; stir frequently. Centrifuge, discard the solution, and wash the precipitate three times with hot water to remove sulfate. Add a few drops of 6 M HNO_3 to the residue and, after reaction ceases, dilute with several drops of water. Centrifuge and withdraw the solution; add it to the acid solution of the sample for the general analysis (Sec. 14.7). If a residue remains, repeat the treatment with sodium carbonate.

(b) The residue from the transposition of the sulfates may be the silver salts, calcium fluoride, oxides, silicates, C, or S. Add 1 ml of water, one drop of 18 M H_2SO_4, and some granules of zinc to reduce the silver halides to free silver. Warm for several minutes, break apart clumps, and disperse the residue with a rod. Centrifuge and discard the solution. Wash the precipitate twice. Add several drops of 6 M HNO_3 to dissolve silver and excess zinc. Identify silver in the solution by precipitating the chloride, dissolving it in ammonia, and reprecipitating the chloride with nitric acid as in Outline 1.

If a residue still remains, it must be decomposed by fusion with sodium and potassium carbonates. (See *Qualitative Analysis and Electrolytic Solutions* by E. J. King, page 571, for detailed directions.)

Appendixes

Appendix A.1

Summary of Solubilities

Class A. Almost all salts soluble in water

1. *Nitrates.* All are soluble but certain basic nitrates; for example, those of bismuth(III) and mercury(II) are only soluble in acids.

2. *Nitrites.* Silver nitrite is sparingly soluble in water, soluble in dilute nitric acid.

3. *Acetates.* The acetates of silver, mercury(I), and chromium(II) are sparingly soluble in water; basic acetates such as that of iron(III) are insoluble. All dissolve in dilute nitric acid.

Class B. Relatively small number of salts insoluble in water

1. *Chlorides, bromides, iodides, and thiocyanates.* Those of silver, mercury(I), and lead(II), and also HgI_2 and $Hg(SCN)_2$, are insoluble. The salts of lead are the most soluble of these. All are insoluble in dilute acids. Basic salts such as those of bismuth(III), antimony(III), and mercury(II) are soluble in acids.

2. *Sulfates.* The sulfates of barium, strontium, and lead are insoluble; those of silver, calcium, and mercury(I) are sparingly soluble. All are only slightly more soluble in acids. The ignited or dehydrated sulfates of iron(III), aluminum, and chromium(III) dissolve in water or acids with difficulty. Basic sulfates are soluble in acids.

Class C. Almost all compounds insoluble in water

The sodium, potassium, and ammonium salts of these are soluble; other exceptions are noted for each class of substance.

1. *Fluorides.* Somewhat soluble in acids; calcium fluoride is least soluble. Soluble salts are those of silver, aluminum, mercury, and tin.

2. *Sulfides.* All but HgS are soluble in nitric acid; many are soluble in hydrochloric acid. Aqua regia is used to dissolve HgS.

3. *Sulfites, phosphates, arsenates, arsenites, carbonates, and borates.* Insoluble in water but usually soluble in dilute acids. Some phosphates found in nature are attacked with difficulty.

4. *Oxalates.* These are soluble in strong acids. Iron(III) oxalate is soluble; magnesium oxalate sparingly soluble in water.

5. *Chromates.* These are soluble in acids, but lead(II) chromate which has been ignited or fused is not attacked by acids. The chromates of magnesium, calcium, and copper(II) are soluble in water, and strontium chromate is sparingly soluble. Dichromates are usually more soluble than chromates.

6. *Oxides and hydroxides.* The oxides of barium, strontium, and calcium are moderately soluble in water. Most metal oxides are soluble in acids, preferably hydrochloric acid. Arsenic(III) oxide is best dissolved with sodium hydroxide. The oxides of aluminum, chromium(III), and tin(IV) in some forms are virtually insoluble in acids.

Appendix A.2

Solubility Products at 25°C

The values tabulated are for the inactive or crystalline solids.

Substance	Formula	K_{sp}
Aluminum hydroxide	$Al(OH)_3$	1.4×10^{-34}
Barium carbonate	$BaCO_3$	1.6×10^{-9}
Barium chromate	$BaCrO_4$	1.2×10^{-10}
Barium fluoride	BaF_2	2.4×10^{-5}
Barium oxalate	$BaC_2O_4 \cdot H_2O$	1.5×10^{-8}
Barium sulfate	$BaSO_4$	1.0×10^{-10}
Bismuth sulfide	Bi_2S_3	1×10^{-96}
Calcium carbonate	$CaCO_3$	4.7×10^{-9}
Calcium chromate	$CaCrO_4$	7.1×10^{-4}
Calcium fluoride	CaF_2	1.7×10^{-10}
Calcium oxalate	$CaC_2O_4 \cdot H_2O$	2.1×10^{-9}
Calcium sulfate	$CaSO_4 \cdot 2H_2O$	2.4×10^{-5}
Cadmium hydroxide	$Cd(OH)_2$	2.8×10^{-14}
Cadmium oxalate	CdC_2O_4	2.8×10^{-8}
Cadmium sulfide	CdS	7×10^{-27}
Chromium(III) hydroxide	$Cr(OH)_3$	7×10^{-31}
Cobalt(II) hydroxide	$Co(OH)_2$	2×10^{-16}
Cobalt(III) hydroxide	$Co(OH)_3$	1×10^{-43}
Cobalt(II) sulfide	CoS	8×10^{-23}
Copper(II) hydroxide	$Cu(OH)_2$	2.2×10^{-20}
Copper(II) sulfide	CuS	8×10^{-36}
Iron(II) hydroxide	$Fe(OH)_2$	8×10^{-16}
Iron(III) hydroxide	$Fe(OH)_3$	6×10^{-38}
Iron(II) sulfide	FeS	5×10^{-18}
Lead(II) bromide	$PbBr_2$	4.6×10^{-6}
Lead(II) carbonate	$PbCO_3$	1.5×10^{-13}
Lead(II) chloride	$PbCl_2$	1.6×10^{-5}
Lead(II) chromate	$PbCrO_4$	2×10^{-16}
Lead(II) fluoride	PbF_2	2.7×10^{-8}
Lead(II) hydroxide	$Pb(OH)_2$	4×10^{-15}
Lead(II) iodide	PbI_2	7.1×10^{-9}
Lead(II) oxalate	PbC_2O_4	8×10^{-12}
Lead(II) sulfate	$PbSO_4$	1.7×10^{-8}
Lead(II) sulfide	PbS	8×10^{-28}
Magnesium ammonium phosphate*	$MgNH_4PO_4 \cdot 6H_2O$	7.1×10^{-14}
Magnesium carbonate	$MgCO_3$	1×10^{-5}
Magnesium fluoride	MgF_2	6.4×10^{-9}
Magnesium hydroxide	$Mg(OH)_2$	1.1×10^{-11}
Magnesium oxalate	MgC_2O_4	8.6×10^{-5}
Manganese(II) carbonate	$MnCO_3$	8.8×10^{-11}

Substance	Formula	K_{sp}
Manganese(II) hydroxide	$Mn(OH)_2$	1.6×10^{-13}
Manganese(II) sulfide	MnS	1×10^{-11}
Mercury(I) chloride[†]	Hg_2Cl_2	1.3×10^{-18}
Mercury(I) iodide[†]	Hg_2I_2	4×10^{-29}
Mercury(II) oxide[‡]	HgO	3×10^{-26}
Mercury(I) sulfate[†]	Hg_2SO_4	6.8×10^{-7}
Mercury(II) sulfide (black)	HgS	3×10^{-52}
Nickel(II) hydroxide	$Ni(OH)_2$	2×10^{-15}
Nickel(II) sulfide	NiS	2×10^{-21}
Silver acetate	$AgOAc$	2.3×10^{-3}
Silver arsenate	Ag_3AsO_4	1×10^{-22}
Silver bromide	$AgBr$	5.2×10^{-13}
Silver carbonate	Ag_2CO_3	8.2×10^{-12}
Silver chloride	$AgCl$	1.8×10^{-10}
Silver chromate	Ag_2CrO_4	2.4×10^{-12}
Silver cyanide	$AgCN$	2×10^{-16}
Silver iodide	AgI	8.3×10^{-17}
Silver oxide[‡]	Ag_2O	2.6×10^{-8}
Silver phosphate	Ag_3PO_4	1×10^{-21}
Silver sulfate	Ag_2SO_4	1.7×10^{-5}
Silver sulfide	Ag_2S	7×10^{-50}
Silver thiocyanate	$AgSCN$	1.0×10^{-12}
Strontium carbonate	$SrCO_3$	7×10^{-10}
Strontium chromate	$SrCrO_4$	2.2×10^{-6}
Strontium fluoride	SrF_2	7.9×10^{-10}
Strontium oxalate	$SrC_2O_4 \cdot H_2O$	5.6×10^{-8}
Strontium sulfate	$SrSO_4$	7.6×10^{-7}
Tin(II) hydroxide	$Sn(OH)_2$	1.6×10^{-27}
Tin(II) sulfide	SnS	1.0×10^{-25}
Zinc carbonate	$ZnCO_3$	2.1×10^{-11}
Zinc hydroxide	$Zn(OH)_2$	7×10^{-18}
Zinc oxalate	ZnC_2O_4	2.5×10^{-9}
Zinc sulfide	ZnS	8×10^{-25}

* $K_{sp} = [Mg^{++}][NH_4^+][PO_4^{3-}]$.
[†] The mercury(I) ion is Hg_2^{++}; hence $K_{sp} = [Hg_2^{++}][Cl^-]^2$, etc.
[‡] The oxide ion is hydrolyzed to hydroxide; hence $K_{sp} = [Hg^{++}][OH^-]^2$ and $K_{sp} = [Ag^+][OH^-]$.

Appendix A.3

Acidity and Basicity Constants at 25°C*

Name of acid	Formula of acid	K_a	Formula of conjugate base	K_b
Acetic acid	CH_3CO_2H	1.8×10^{-5}	$CH_3CO_2^-$	5.7×10^{-10}
Ammonium ion	NH_4^+	5.7×10^{-10}	NH_3	1.8×10^{-5}
Arsenic acid	H_3AsO_4	6.0×10^{-3}	$H_2AsO_4^-$	1.7×10^{-12}
	$H_2AsO_4^-$	1.0×10^{-7}	$HAsO_4^{--}$	1.0×10^{-7}
	$HAsO_4^{--}$	4×10^{-12}	AsO_4^{3-}	2.5×10^{-3}
Benzoic acid	$C_6H_5CO_2H$	6.3×10^{-5}	$C_6H_5CO_2^-$	1.6×10^{-10}
Boric acid	H_3BO_3	5.8×10^{-10}	$B(OH)_4^-$	1.7×10^{-5}
Carbonic acid	$H_2CO_3 + CO_2$	4.4×10^{-7}	HCO_3^-	2.2×10^{-8}
	HCO_3^-	4.7×10^{-11}	CO_3^{--}	2.1×10^{-4}
Chromic acid	H_2CrO_4	strong	$HCrO_4^-$	—
	$HCrO_4^-$	3.0×10^{-7}	CrO_4^{--}	3.3×10^{-8}
Formic acid	HCO_2H	1.8×10^{-4}	HCO_2^-	5.7×10^{-11}
Hydrocyanic acid	HCN	6×10^{-10}	CN^-	1.7×10^{-5}
Hydrofluoric acid	HF	6.7×10^{-4}	F^-	1.5×10^{-11}
Hydrosulfuric acid[†]	H_2S	1.0×10^{-7}	HS^-	1.0×10^{-7}
	HS^-	1×10^{-14}	S^{--}	1
Monochloracetic acid	$ClCH_2CO_2H$	1.4×10^{-3}	$ClCH_2CO_2^-$	7.0×10^{-12}
Nitrous acid	HNO_2	5.1×10^{-4}	NO_2^-	2.0×10^{-11}
Oxalic acid	$H_2C_2O_4$	5.4×10^{-2}	$HC_2O_4^-$	1.9×10^{-13}
	$HC_2O_4^-$	5.4×10^{-5}	$C_2O_4^{--}$	1.9×10^{-10}
Phosphoric acid	H_3PO_4	7.1×10^{-3}	$H_2PO_4^-$	1.4×10^{-12}
	$H_2PO_4^-$	6.3×10^{-8}	HPO_4^{--}	1.6×10^{-7}
	HPO_4^{--}	4.2×10^{-13}	PO_4^{3-}	2.4×10^{-2}
Propionic acid	$CH_3CH_2CO_2H$	1.3×10^{-5}	$CH_3CH_2CO_2^-$	7.4×10^{-10}
Sulfamic acid	HNH_2SO_3	1.0×10^{-1}	$NH_2SO_3^-$	1.0×10^{-13}
Sulfuric acid	H_2SO_4	strong	HSO_4^-	—
	HSO_4^-	1.0×10^{-2}	SO_4^{--}	1.0×10^{-12}
Sulfurous acid	$H_2SO_3 + SO_2$	1.7×10^{-2}	HSO_3^-	6.4×10^{-13}
	HSO_3^-	6.2×10^{-8}	SO_3^{--}	1.6×10^{-7}
Triethanol-ammonium ion	$(C_2H_4OH)_3NH^+$	1.7×10^{-8}	$(C_2H_4OH)_3N$	5.9×10^{-7}

* For concentrations on the molarity scale.
† Although K_a for H_2S is well established, K_a for HS^- is highly uncertain, with reported values ranging from 10^{-13} to 10^{-17}. The value tabulated here may be too high but is consistent with solubility products of the sulfides in Appendix A.2.

Appendix A.4

The Ion Product of Water at Various Temperatures

Temperature, °C	K_w, M^2	pK_w
0	0.11×10^{-14}	14.94
20	0.68×10^{-14}	14.17
25	1.00×10^{-14}	14.00
30	1.47×10^{-14}	13.83
37*	2.40×10^{-14}	13.62
90	3.78×10^{-13}	12.42
100	5.51×10^{-13}	12.26
110	7.48×10^{-13}	12.13

* Normal body temperature

Appendix A.5

The Ion Product of Hydrogen Sulfide*

For solutions saturated with H_2S at 1 atm pressure (Sec. 2.6)

$$K_{ip} = [H^+]^2[S^{--}]$$

Temperature, °C	K_{ip}, M^3
25	1.0×10^{-22}
100	1.1×10^{-18}

* These values are uncertain insofar as K_a of HS^- is uncertain; see footnote to Appendix A.3.

Appendix B.1

Reagent Solutions

Reagent	Use	Bottle	Concn.	Method of preparation
*Acid, acetic	B	30	6 M	Dilute the concd (17 M) acid
*Acid, hydrochloric	C	30	2 M	Dilute the concd (12 M) acid
*Acid, hydrochloric	B	30	6 M	Dilute the concd (12 M) acid
*Acid, hydrochloric	B	30gs	12 M	Use the concd reagent
*Acid, nitric	B	30	6 M	Dilute the concd (16 M) acid
*Acid, nitric	B	30gs	16 M	Use the concd reagent
*Acid, sulfuric	A	8	6 M	Dilute the concd (18 M) acid
*Acid, sulfuric	B	30gs	18 M	Use the concd reagent
*Aluminon	C	8	0.1%	1.0 g/l of ammonium salt of aurintricarboxylic acid
*Ammonia	B	30	6 M	Dilute the concd (15 M) base
*Ammonia	B	30	15 M	Use the concd reagent
*Ammonium acetate	C	8	3 M	231.2 g/l
*Ammonium carbonate	C	8†		See B.2
*Ammonium chloride	C	8	satd	321 g/l. Approx. 6 M
*Ammonium molybdate	A	8		See B.2
*Ammonium nitrate	C	8†	0.2 M	16.0 g/l
*Ammonium oxalate	C	8†	0.2 M	28.4 g $(NH_4)_2C_2O_4 \cdot H_2O$ per liter
*Ammonium sulfate	C	8†	1 M	132 g/l
*Ammonium thiocyanate	C	8	satd	See B.2
*Barium chloride	B	8	0.2 M	48.8 g $BaCl_2 \cdot 2H_2O$ per liter. Store in bottle of borosilicate glass.
*Barium hydroxide	A	8	0.2 M	Satd soln (approx 63 g $Ba(OH)_2 \cdot 8H_2O$ per liter)
*Barium nitrate	A	8	0.1 M	26.1 g/l
*Calcium nitrate	A	8	0.1 M	23.6 g $Ca(NO_3)_2 \cdot 4H_2O$ per liter
*Dimethylglyoxime	C	8	1%	10 g per liter of 95% ethyl alcohol
*Hydrogen peroxide	B	8am	3%	Use USP solution
*Iron(III) chloride	A	8	0.33 M	90 g $FeCl_3 \cdot 6H_2O$ to 20 ml 6 M HCl. Dil to 1 liter
*Lead acetate	C	8†	0.2 M	76.0 g $Pb(OAc)_2 \cdot 3H_2O$ per liter
*Lead nitrate	A	8†	0.1 M	33.1 g/l
*Magnesia mixture	C	8		See B.2
*Magnesium uranyl acetate	C	8		See B.2
*Manganese(II) chloride	A	8	satd	Sat 12 M HCl with $MnCl_2 \cdot 4H_2O$
*Mercury(II) chloride	C	8†	0.1 M	27.2 g/l
*Potassium chromate	C	8	0.5 M	97.1 g/l
*Potassium ferrocyanide	C†	8	0.2 M	84.5 g $K_4Fe(CN)_6 \cdot 3H_2O$ per liter
*Potassium hydroxide	C	30p	0.5 M	28.0 g/l
*Potassium permanganate	A	8am	0.02 M	3.2 g/l

Reagent	Use	Bottle	Concn.	Method of preparation
*Silver nitrate	B	8am	0.2 M	34.0 g/l
*S. and O. reagent	C†	8	0.05%	0.500 g p-nitrobenzeneazoresorcinol in 1 liter of 0.025 M NaOH
Sodium cobaltinitrite	C	SS	0.33 M	135 g/l. Keep cool
*Disodium hydrogen phosphate	C	8	0.5 M	Add 71.0 g anhyd Na_2HPO_4 gradually to water. Dil to 1 liter
*Sodium hydroxide	B	30p	6 M	240 g/l
Solution A	A†	SSam	3.3 M	170 g $AgNO_3$ in 300 ml H_2O
Solution B	A†	SS		25 ml satd $(NH_4)_2CO_3$ + 10 ml 15 M NH_3 + 100 ml H_2O
*Thioacetamide	B	30	13%	See B.2
*Tin(II) chloride	C	8		See B.2
*Triethanolamine	C†	8	20%	20% by volume with water

Key: Solutions that are suggested for the reagent kit are marked with an asterisk. Under "Use," A indicates solutions used only for the anion analysis, B those used in both anion and cation analysis, and C those used only in cation analysis. Reagents not required for the abbreviated procedures of Chapters 4 and 6 are marked with †. Under "Bottle," 8 is for the 8-ml size, 30 for the 30-ml size, SS for side shelf reagent (250 ml); am stands for amber bottle, gs for glass stoppered, and p for polyethylene. Although 8 ml is an ample supply of most reagents, stock solutions should be available for refills. Like the side shelf reagents, they may be kept in 250-ml bottles fitted with droppers. Reference is made to B.2 for more detailed directions or comments on certain solutions. Unless otherwise indicated, the weight in the last column of the table is dissolved in distilled water and made up to a final volume of 1 liter.

Appendix B.2

Solutions. Directions and Comments

Ammonium carbonate: Dissolve 192 g of powered reagent in 500 ml of water and 80 ml of 15 M NH_3. Dilute to 1 liter.

Ammonium molybdate: Dissolve 90 g of $(NH_4)_6Mo_7O_{24} \cdot 4H_2O$ in 100 ml of 6 M NH_3. Add 240 g of NH_4NO_3, and after it has dissolved dilute to 1 liter.

Ammonium thiocyanate: Saturate a mixture of equal volumes of ethyl ether and isoamyl alcohol with solid NH_4SCN.

Magnesia mixture: Dissolve 130 g of $Mg(NO_3)_2 \cdot 6H_2O$ and 240 g of NH_4NO_3 in 500 ml water. Add 150 ml of 15 M NH_3 and dilute to 1 liter. Store in a polyethylene bottle.

Magnesium uranyl acetate: Dissolve 30 g of $UO_2(OAc)_2 \cdot 2H_2O$ in a mixture of 120 ml of water and 100 ml of glacial acetic acid. Prepare also a solution of magnesium acetate by adding slowly 148.5 g of $Mg(OAc)_2 \cdot 4H_2O$ to a hot mixture of 40 ml of water and 320 ml of glacial acetic acid. Mix the magnesium and uranyl solutions while they are still warm, and allow the mixture to stand overnight. Filter through cotton or sintered glass.

Thioacetamide: Dissolve 130 g in a liter; the solution is almost saturated. After being kept for more than a year in a clear glass bottle, the solution is still effective in precipitating sulfides. Decomposition occurs on standing, as shown by the precipitation of sulfur; this does no harm. More serious is the gradual formation of sulfate. If the reagent gives a precipitate with barium chloride, a fresh solution should be prepared.

Tin(II) chloride: Let 45.0 g of $SnCl_2 \cdot H_2O$ stand with 170 ml of 12 M HCl until the lumps disintegrate. Dilute slowly to 1 liter. Keep tin shot in the solution.

Dithizone paper: Dissolve 0.1 g of dithizone (diphenylthiocarbazone) in 100 ml of acetone. Transfer the solution to a cylinder. Cut Whatman Spot Reaction Paper (No. 120) into strips about 1 cm wide and dip the strips in the solution. Spread them over glass rods in the hood to dry. Then cut them into squares. If stored in a screw-capped vial, the squares will keep for 6 months or more.

Appendix B.3

Solid and Pure Liquid Reagents

Reagent	Use	Reagent	Use
*Aluminum wire, 5-mm lengths	C†	*Potassium dichromate	A
*Ammonium thiocyanate	C	*Potassium ferricyanide	B
Barium chloride	C	*Sodium bismuthate	C
Calcium chloride	C	*Sodium carbonate	B
Carbon tetrachloride	A	Sodium chloride	C
Carborundum chips (boiling		*Sodium dithionite	C†
stones)	A	*Sodium fluoride	B
Cotton	B	*Sodium nitrite	B
*Devarda's alloy (Al can be used		Sodium sulfate	A
instead)	A	Sodium sulfite	A
Ethyl alcohol, 95%	B	Strontium chloride	C
*Oxalic acid	C†	*Sulfamic acid	B
*Potassium chlorate	C	*Zinc, 20 mesh	B
Potassium chloride	C	Zinc sulfide (or sodium sulfide)	A

Key: See B.1 for meaning of symbols under "Use." Asterisks mark reagents that can be issued in the reagent kit. Vials or 8-ml bottles with screw caps can be used.

Appendix B.4

Test Papers

    Dithizone paper (B.2)
    Indicator paper, wide-range, such as, Hydrion Vivid 1–11
    Indicator paper, short-range, such as, Hydrion pH 0.0–1.5
    Litmus paper, red and blue
    Lead acetate paper

Appendix B.5

Standard Solutions of the Ions

(10 mg of ion per ml)

		CATIONS	
Ion	*Compound*	*Grams/liter*	*Solvent*
Ag^+	$AgNO_3$	15.8	Water
Hg_2^{++}	$Hg_2(NO_3)_2 \cdot 2H_2O$	14.0	0.6 M HNO_3
Pb^{++}	$Pb(NO_3)_2$	16.0	Water
Hg^{++}	$Hg(NO_3)_2 \cdot \frac{1}{2}H_2O$	16.7	0.16 M HNO_3
Bi^{3+}	$Bi(NO_3)_3 \cdot 5H_2O$	23.2	3.0 M HNO_3
Cu^{++}	$Cu(NO_3)_2 \cdot 3H_2O$	38.0	Water
Cd^{++}	$Cd(NO_3)_2 \cdot 4H_2O$	27.5	Water
As^{III}	As_2O_3	13.2	4 M HCl
As^V	$Na_2HAsO_4 \cdot 7H_2O$	41.7	Water
Sb^{III}	$SbCl_3$	18.8	2.8 M HCl
Sn^{++}	$SnCl_2 \cdot 2H_2O$	19.0	2.8 M HCl
Sn^{IV}	$SnCl_4 \cdot 5H_2O$	29.6	2.4 M HCl
Al^{3+}	$Al(NO_3)_3 \cdot 9H_2O$	139.0	0.024 M HNO_3
Cr^{3+}	$Cr(NO_3)_3 \cdot 9H_2O$	77.0	0.024 M HNO_3
Fe^{3+}	$Fe(NO_3)_3 \cdot 9H_2O$	72.4	0.024 M HNO_3
Co^{++}	$Co(NO_3)_2 \cdot 6H_2O$	49.5	Water
Ni^{++}	$Ni(NO_3)_2 \cdot 6H_2O$	49.5	Water
Mn^{++}	$Mn(NO_3)_2, 50\%$	42.4 ml	Water
Zn^{++}	$Zn(NO_3)_2 \cdot 6H_2O$	45.5	Water
Ca^{++}	$Ca(NO_3)_2 \cdot 4H_2O$	59.0	Water
Sr^{++}	$Sr(NO_3)_2 \cdot 4H_2O$	32.4	Water
Ba^{++}	$Ba(NO_3)_2$	19.0	Water
Mg^{++}	$Mg(NO_3)_2 \cdot 6H_2O$	106.0	Water
Na^+	$NaNO_3$	37.0	Water
K^+	KNO_3	25.9	Water
NH_4^+	NH_4NO_3	44.5	Water

	ANIONS	
Ion	*Compound*	*Grams/liter*
F^-	NaF	22.1
Cl^-	NaCl	16.5
Br^-	KBr	14.9
I^-	KI	13.1
S^{--}	$Na_2S \cdot 9H_2O$	75.2
SO_3^{--}	Na_2SO_3	15.7
SO_4^{--}	$(NH_4)_2SO_4$	13.8
NO_2^-	$NaNO_2$	15.0
NO_3^-	KNO_3	16.3
PO_4^{3-}	$Na_3PO_4 \bullet 12H_2O$	39.6
CO_3^{--}	Na_2CO_3	17.6

Appendix C

Apparatus

The laboratory should be equipped with a centrifuge for every five or six students and should have one or two triple-beam balances sensitive to 0.01 g. The following list of apparatus for student desks can be modified to suit the circumstances of the laboratory and the amount of laboratory work to be covered. Items marked with an asterisk are for the reagent kit, which can be shared by several students working different periods.

3 beakers, 250-ml
1 each beakers, 5-, 10-, 50-, 100-, 400-ml
*14 bottles, 8-ml, with screw cap
*28 bottles, 8-ml, with droppers
*3 bottles, 8-ml, amber, with droppers
*7 bottles, 30-ml, with droppers
*2 bottles, 30-ml, polyethylene, with droppers
*3 bottles, 30-ml, glass-stoppered
1 casserole, 15-ml
8 centrifuge tubes, 3-ml
1 crucible, 1.5-ml
1 flask, Erlenmeyer, 25-ml
2 glass squares, blue
1 graduated cylinder, 10-ml
1 medicine dropper
1 gas lighter or box of safety matches
1 Nichrome wire with cork handle
1 ring
1 ring stand
1 spatula, monel or nickel
1 screw clamp
1 test tube holder, wire
1 tongs
1 tripod
1 water bath cover
1 wing top

2 wire gauzes, with asbestos centers
1 wood block for test tubes
1 wood block for specimen vials
*1 wood rack for 8-ml bottles
*1 wood tray for 30-ml bottles
1 asbestos mat
4 pasteur pipets, with 50-mm capillary tips
2 specimen vials
1 spot plate, white
12 test tubes, 10×75 mm
2 watch glasses, 25 mm
1 wash bottle, polyethylene, 16-oz
1 sponge
1 towel
1 burner, micro, with rubber tube
1 burner, bunsen, with rubber tube
1 file or glass scorer
1 forceps
1 brush, tapered
12 corks, No. 1
1 box labels
1 pkg pipe cleaners
12 rubber bands, $1\frac{1}{4}$ in. long
4 rubber bulbs for pipets
4 ft glass rod, 2-mm
1 vial each of red and blue litmus, lead acetate paper

Appendix D

Four-Place Logarithms

N	0	1	2	3	4	5	6	7	8	9
1.0	.0000	.0043	.0086	.0128	.0170	.0212	.0253	.0294	.0334	.0374
1.1	.0414	.0453	.0492	.0531	.0569	.0607	.0645	.0682	.0719	.0756
1.2	.0792	.0828	.0864	.0899	.0934	.0969	.1004	.1038	.1072	.1106
1.3	.1139	.1173	.1206	.1239	.1271	.1303	.1335	.1367	.1399	.1430
1.4	.1461	.1492	.1523	.1553	.1584	.1614	.1644	.1673	.1703	.1732
1.5	.1761	.1790	.1818	.1847	.1875	.1903	.1931	.1959	.1987	.2014
1.6	.2041	.2068	.2095	.2122	.2148	.2175	.2201	.2227	.2253	.2279
1.7	.2305	.2330	.2355	.2381	.2406	.2430	.2455	.2480	.2504	.2529
1.8	.2553	.2577	.2601	.2625	.2648	.2672	.2695	.2718	.2742	.2765
1.9	.2788	.2810	.2833	.2856	.2878	.2900	.2923	.2945	.2967	.2989
2.0	.3010	.3032	.3054	.3075	.3096	.3118	.3139	.3160	.3181	.3202
2.1	.3222	.3243	.3263	.3284	.3304	.3324	.3345	.3365	.3385	.3404
2.2	.3424	.3444	.3464	.3483	.3503	.3522	.3541	.3560	.3579	.3598
2.3	.3617	.3636	.3655	.3674	.3692	.3711	.3729	.3748	.3766	.3784
2.4	.3802	.3820	.3838	.3856	.3874	.3892	.3909	.3927	.3945	.3962
2.5	.3979	.3997	.4014	.4031	.4048	.4065	.4082	.4099	.4116	.4133
2.6	.4150	.4166	.4183	.4200	.4216	.4233	.4249	.4265	.4281	.4298
2.7	.4314	.4330	.4344	.4362	.4378	.4393	.4409	.4425	.4440	.4456
2.8	.4472	.4487	.4503	.4518	.4533	.4548	.4564	.4579	.4594	.4609
2.9	.4624	.4639	.4654	.4669	.4684	.4698	.4713	.4728	.4742	.4757
3.0	.4771	.4786	.4800	.4814	.4829	.4843	.4857	.4871	.4886	.4900
3.1	.4914	.4928	.4942	.4955	.4969	.4983	.4997	.5011	.5024	.5038
3.2	.5052	.5065	.5079	.5092	.5106	.5119	.5132	.5146	.5159	.5172
3.3	.5185	.5198	.5211	.5224	.5238	.5250	.5263	.5276	.5289	.5302
3.4	.5315	.5328	.5340	.5353	.5366	.5378	.5391	.5403	.5416	.5428
3.5	.5441	.5453	.5465	.5478	.5490	.5502	.5515	.5527	.5539	.5551
3.6	.5563	.5575	.5587	.5599	.5611	.5623	.5635	.5647	.5659	.5670
3.7	.5682	.5694	.5705	.5717	.5729	.5740	.5752	.5763	.5775	.5786
3.8	.5798	.5809	.5821	.5832	.5843	.5855	.5866	.5877	.5888	.5900
3.9	.5911	.5922	.5933	.5944	.5955	.5966	.5977	.5988	.5999	.6010
4.0	.6021	.6031	.6042	.6053	.6064	.6075	.6085	.6096	.6107	.6117
4.1	.6128	.6138	.6149	.6160	.6170	.6181	.6191	.6201	.6212	.6222
4.2	.6233	.6243	.6253	.6263	.6274	.6284	.6294	.6304	.6314	.6325
4.3	.6335	.6345	.6355	.6365	.6375	.6385	.6395	.6405	.6415	.6425
4.4	.6434	.6444	.6454	.6464	.6474	.6484	.6493	.6503	.6513	.6523
4.5	.6532	.6542	.6551	.6561	.6571	.6580	.6590	.6599	.6609	.6618
4.6	.6628	.6637	.6646	.6656	.6665	.6675	.6684	.6693	.6703	.6712
4.7	.6721	.6730	.6740	.6749	.6758	.6767	.6776	.6785	.6794	.6803
4.8	.6812	.6822	.6831	.6840	.6849	.6857	.6866	.6875	.6884	.6893
4.9	.6902	.6911	.6920	.6929	.6937	.6946	.6955	.6964	.6972	.6981
5.0	.6990	.6998	.7007	.7016	.7024	.7033	.7042	.7050	.7059	.7067
5.1	.7076	.7084	.7093	.7101	.7110	.7118	.7127	.7135	.7143	.7152
5.2	.7160	.7168	.7177	.7185	.7193	.7202	.7210	.7218	.7226	.7235
5.3	.7243	.7251	.7259	.7267	.7275	.7284	.7292	.7300	.7309	.7316
5.4	.7324	.7332	.7340	.7348	.7356	.7364	.7370	.7380	.7388	.7396
N	0	1	2	3	4	5	6	7	8	9

N	0	1	2	3	4	5	6	7	8	9
5.5	.7404	.7412	.7419	.7427	.7435	.7443	.7451	.7459	.7466	.7474
5.6	.7482	.7490	.7497	.7505	.7513	.7521	.7528	.7536	.7544	.7551
5.7	.7559	.7566	.7574	.7582	.7589	.7597	.7604	.7612	.7619	.7627
5.8	.7634	.7642	.7649	.7657	.7664	.7672	.7679	.7686	.7694	.7701
5.9	.7709	.7716	.7723	.7731	.7738	.7745	.7753	.7760	.7767	.7774
6.0	.7782	.7789	.7796	.7803	.7810	.7818	.7825	.7832	.7839	.7846
6.1	.7853	.7860	.7868	.7875	.7882	.7889	.7896	.7903	.7910	.7917
6.2	.7924	.7931	.7938	.7945	.7952	.7959	.7966	.7973	.7980	.7987
6.3	.7993	.8000	.8007	.8014	.8021	.8028	.8035	.8041	.8048	.8055
6.4	.8062	.8069	.8075	.8082	.8089	.8096	.8102	.8109	.8116	.8122
6.5	.8129	.8136	.8143	.8149	.8156	.8162	.8169	.8176	.8182	.8189
6.6	.8195	.8202	.8209	.8215	.8222	.8228	.8235	.8241	.8248	.8254
6.7	.8261	.8267	.8274	.8280	.8287	.8293	.8300	.8306	.8312	.8319
6.8	.8325	.8332	.8338	.8344	.8351	.8357	.8363	.8370	.8376	.8382
6.9	.8389	.8395	.8401	.8407	.8414	.8420	.8426	.8432	.8439	.8445
7.0	.8451	.8457	.8463	.8470	.8476	.8482	.8488	.8494	.8500	.8507
7.1	.8513	.8519	.8525	.8531	.8537	.8543	.8549	.8555	.8561	.8567
7.2	.8573	.8579	.8585	.8591	.8597	.8603	.8609	.8615	.8621	.8627
7.3	.8633	.8639	.8645	.8651	.8657	.8663	.8667	.8675	.8681	.8686
7.4	.8692	.8698	.8704	.8710	.8716	.8722	.8727	.8733	.8739	.8745
7.5	.8751	.8756	.8762	.8768	.8774	.8780	.8785	.8791	.8797	.8802
7.6	.8808	.8814	.8820	.8825	.8831	.8837	.8842	.8848	.8854	.8859
7.7	.8865	.8871	.8876	.8882	.8887	.8893	.8899	.8904	.8910	.8915
7.8	.8921	.8927	.8932	.8938	.8943	.8949	.8954	.8960	.8965	.8971
7.9	.8976	.8982	.8987	.8993	.8998	.9004	.9009	.9015	.9020	.9026
8.0	.9031	.9036	.9042	.9047	.9053	.9058	.9063	.9069	.9074	.9080
8.1	.9085	.9090	.9096	.9101	.9106	.9112	.9117	.9122	.9128	.9133
8.2	.9138	.9143	.9149	.9154	.9159	.9165	.9170	.9175	.9180	.9186
8.3	.9191	.9196	.9201	.9207	.9212	.9217	.9222	.9227	.9232	.9238
8.4	.9243	.9248	.9253	.9258	.9263	.9267	.9274	.9279	.9284	.9289
8.5	.9294	.9299	.9304	.9310	.9315	.9320	.9325	.9330	.9335	.9340
8.6	.9345	.9350	.9355	.9360	.9365	.9370	.9375	.9380	.9385	.9390
8.7	.9395	.9400	.9405	.9410	.9415	.9420	.9425	.9430	.9435	.9440
8.8	.9445	.9450	.9455	.9460	.9465	.9470	.9474	.9479	.9484	.9489
8.9	.9494	.9499	.9504	.9509	.9513	.9518	.9523	.9528	.9533	.9538
9.0	.9542	.9547	.9552	.9557	.9562	.9567	.9571	.9576	.9581	.9586
9.1	.9590	.9595	.9600	.9605	.9610	.9614	.9619	.9624	.9628	.9633
9.2	.9638	.9643	.9647	.9652	.9657	.9661	.9666	.9671	.9676	.9680
9.3	.9685	.9690	.9694	.9699	.9704	.9708	.9713	.9717	.9722	.9727
9.4	.9731	.9736	.9741	.9745	.9750	.9754	.9759	.9764	.9768	.9773
9.5	.9777	.9782	.9786	.9791	.9796	.9800	.9805	.9809	.9814	.9818
9.6	.9823	.9827	.9832	.9836	.9841	.9845	.9850	.9854	.9859	.9863
9.7	.9868	.9872	.9877	.9881	.9886	.9890	.9895	.9900	.9903	.9908
9.8	.9912	.9917	.9921	.9926	.9930	.9934	.9939	.9943	.9948	.9952
9.9	.9956	.9961	.9965	.9970	.9974	.9978	.9983	.9987	.9991	.9996
N	0	1	2	3	4	5	6	7	8	9

Table of Atomic Weights

1969 Values based on Carbon-12*

Element	Symbol	Atomic number	Atomic weight	Element	Symbol	Atomic number	Atomic weight
Actinium	Ac	89	(227)	Iodine	I	53	126.9045
Aluminum	Al	13	26.9815	Iridium	Ir	77	192.2_2
Americium	Am	95	(241, 243)	Iron	Fe	26	55.84_7
Antimony	Sb	51	121.7_5	Krypton	Kr	36	83.80
Argon	Ar	18	39.94_8	Lanthanum	La	57	138.905_5
Arsenic	As	33	74.9216	Lawrencium	Lr	103	(256)
Astatine	At	85	(210)	Lead	Pb	82	207.2
Barium	Ba	56	137.3_4	Lithium	Li	3	6.941
Berkelium	Bk	97	(247, 249)	Lutetium	Lu	71	174.97
Beryllium	Be	4	9.01218	Magnesium	Mg	12	24.305
Bismuth	Bi	83	208.9806	Manganese	Mn	25	54.9380
Boron	B	5	10.81	Mendelevium	Md	101	(257, 258)
Bromine	Br	35	79.904	Mercury	Hg	80	200.5_9
Cadmium	Cd	48	112.40	Molybdenum	Mo	42	95.9_4
Calcium	Ca	20	40.08	Neodymium	Nd	60	144.2_4
Californium	Cf	98	(251, 252, 254)	Neon	Ne	10	20.17_9
				Neptunium	Np	93	237.0482
Carbon	C	6	12.011	Nickel	Ni	28	58.7_1
Cerium	Ce	58	140.12	Niobium	Nb	41	92.9064
Cesium	Cs	55	132.9055	Nitrogen	N	7	14.0067
Chlorine	Cl	17	35.453	Nobelium	No	102	(255)
Chromium	Cr	24	51.996	Osmium	Os	76	190.2
Cobalt	Co	27	58.9332	Oxygen	O	8	15.999_4
Copper	Cu	29	63.54_6	Palladium	Pd	46	106.4
Curium	Cm	96	(243–248, 250)	Phosphorus	P	15	30.9738
				Platinum	Pt	78	195.0_9
Dysprosium	Dy	66	162.5_0	Plutonium	Pu	94	(238–242, 244)
Einsteinium	Es	99	(253, 254)				
Erbium	Er	68	167.2_6	Polonium	Po	84	(209, 210)
Europium	Eu	63	151.96	Potassium	K	19	39.10_2
Fermium	Fm	100	(257)	Praseodymium	Pr	59	140.9077
Fluorine	F	9	18.9984	Prometheum	Pm	61	(145, 147)
Francium	Fr	87	(223)	Protactinium	Pa	91	231.0359
Gadolinium	Gd	64	157.2_5	Radium	Ra	88	226.0254
Gallium	Ga	31	69.72	Radon	Rn	86	(222)
Germanium	Ge	32	72.5_9	Rhenium	Re	75	186.2
Gold	Au	79	196.9665	Rhodium	Rh	45	102.9055
Hafnium	Hf	72	178.4_9	Rubidium	Rb	37	85.467_8
Helium	He	2	4.00260	Ruthenium	Ru	44	101.0_7
Holmium	Ho	67	164.9303	Samarium	Sm	62	150.4
Hydrogen	H	1	1.008_0	Scandium	Sc	21	44.9559
Indium	In	49	114.82	Selenium	Se	34	78.9_6

Element	Symbol	Atomic number	Atomic weight	Element	Symbol	Atomic number	Atomic weight
Silicon	Si	14	28.08_6	Thulium	Tm	69	168.9342
Silver	Ag	47	107.868	Tin	Sn	50	118.6_9
Sodium	Na	11	22.9898	Titanium	Ti	22	47.9_0
Strontium	Sr	38	87.62	Tungsten	W	74	183.8_5
Sulfur	S	16	32.06	Uranium	U	92	238.029
Tantalum	Ta	73	180.947_9	Vanadium	V	23	50.941_4
Technetium	Tc	43	98.9062	Xenon	Xe	54	131.30
Tellurium	Te	52	127.6_0	Ytterbium	Yb	70	173.0_4
Terbium	Tb	65	158.9254	Yttrium	Y	39	88.9059
Thallium	Tl	81	204.3_7	Zinc	Zn	30	65.3_7
Thorium	Th	90	232.0381	Zirconium	Zr	40	91.22

* Numbers in parentheses are mass numbers of important radioactive isotopes.

Answers to Numerical Problems

1.10. 2×10^{19}

1.11. $3.6 \times 10^7 \, M$

1.12. 100%

1.13. 8×10^{-15}

1.16. 8.45 kcal

2.10. (b) $9 \times 10^{-9} \, M$ for AgCl

     (c) $3.7 \times 10^{-6} \, M$

     (d) 99.98%

2.11. (b) $6.0 \times 10^{-25} \, M \, Cu^{++}$

2.12. $2 \times 10^{-22} \, M^2$

2.13. $1.5 \times 10^{-5} \, M$

2.14. 17.0 g

2.15. $11 \, M \, Mn^{++}$

2.16. 0.004%

2.17. 0.76

2.18. $0.11 \, M$

2.19. 86

2.20. (a) 1.84 mg

     (b) 1.95 mg

7.15. 7.4×10^{-8} mg/ml

7.16. $0.029 \, M$

7.17. $8 \times 10^{10} \, M$

7.18. (a) $3 \times 10^{15} \, M^{-2}$

     (b) $1.1 \times 10^4 \, M^{-4}$

7.19. (a) 3×10^{-2}

     (b) $1.13 \, M^{-1}$

8.19. $0.3 \, M$

8.20. (b) $0.10 \, M$

8.21. $3.6 \times 10^{-23} \, M^2$

9.17. $0.06 \, M$

9.18. $0.0145 \, M$; $10.6 \, M$

9.19. 0.11 g

9.20. $4 \times 10^{21} \, M^2$

10.13. (a) 8.1

10.14. $4.5 \times 10^{-8} \, M$

10.15. (c) 0.19%

10.16. $1.0 \times 10^{-4} \, M$ at pH 0

11.9. 0.72 mg

11.10. 8×10^{-17} atm^2

Index